Springer
Proceedings in Physics 82

Springer
Tokyo
Berlin
Heidelberg
New York
Barcelona
Budapest
Hong Kong
London
Milan
Paris
Santa Clara
Singapore

Springer Proceedings in Physics

Managing Editor: H. K. V. Lotsch

46 *Cellular Automata and Modeling of Complex Physical Systems*
Editors: P. Manneville, N. Boccara, G. Y. Vichniac, and R. Bidaux

47 *Number Theory and Physics*
Editors: J.-M. Luck, P. Moussa, and M. Waldschmidt

48 *Many-Atom Interactions in Solids*
Editors: R. M. Nieminen, M. J. Puska, and M. J. Manninen

49 *Ultrafast Phenomena in Spectroscopy*
Editors: E. Klose and B. Wilhelmi

50 *Magnetic Properties of Low-Dimensional Systems II: New Developments*
Editors: L. M. Falicov, F. Mejía-Lira, and J. L. Morán-López

51 *The Physics and Chemistry of Organic Superconductors*
Editors: G. Saito and S. Kagoshima

52 *Dynamics and Patterns in Complex Fluids: New Aspects of the Physics – Chemistry Interface*
Editors: A. Onuki and K. Kawasaki

53 *Computer Simulation Studies in Condensed-Matter Physics III*
Editors: D. P. Landau, K. K. Mon, and H.-B. Schüttler

54 *Polycrystalline Semiconductors II*
Editors: J. H. Werner and H. P. Strunk

55 *Nonlinear Dynamics and Quantum Phenomena in Optical Systems*
Editors: R. Vilaseca and R. Corbalán

56 *Amorphous and Crystalline Silicon Carbide III, and Other Group IV-IV Materials*
Editors: G. L. Harris, M. G. Spencer, and C. Y. Yang

57 *Evolutionary Trends in the Physical Sciences*
Editors: M. Suzuki and R. Kubo

58 *New Trends in Nuclear Collective Dynamics*
Editors: Y. Abe, H. Horiuchi, and K. Matsuyanagi

59 *Exotic Atoms in Condensed Matter*
Editors: G. Benedek and H. Schneuwly

60 *The Physics and Chemistry of Oxide Superconductors*
Editors: Y. Iye and H. Yasuoka

61 *Surface X-Ray and Neutron Scattering*
Editors: H. Zabel and I. K. Robinson

62 *Surface Science: Lectures on Basic Concepts and Applications*
Editors: F. A. Ponce and M. Cardona

63 *Coherent Raman Spectroscopy: Recent Advances*
Editors: G. Marowsky and V. V. Smirnov

64 *Superconducting Devices and Their Applications*
Editors: H. Koch and H. Lübbig

65 *Present and Future of High-Energy Physics*
Editors. K.-I. Aoki and M. Kobayashi

66 *The Structure and Conformation of Amphiphilic Membranes*
Editors: R. Lipowsky, D. Richter, and K. Kremer

67 *Nonlinearity with Disorder*
Editors: F. Abdullaev, A. R. Bishop, and S. Pnevmatikos

68 *Time-Resolved Vibrational Spectroscopy V*
Editor: H. Takahashi

69 *Evolution of Dynamical Structures in Complex Systems*
Editors: R. Friedrich and A. Wunderlin

70 *Computational Approaches in Condensed-Matter Physics*
Editors: S.Miyashita,M.Imada, and H.Takayama

71 *Amorphous and Crystalline Silicon Carbide IV*
Editors: C. Y. Yang, M. M. Rahman, and G. L. Harris

72 *Computer Simulation Studies in Condensed-Matter Physics IV*
Editors: D. P. Landau, K. K. Mon, and H.-B. Schüttler

73 *Surface Science: Principles and Applications*
Editors: R. F. Howe, R. N. Lamb, and K. Wandelt

74 *Time-Resolved Vibrational Spectroscopy VI*
Editors: A. Lau, F. Siebert, and W. Werncke

75 *Computer Simulation Studies in Condensed-Matter Physics V*
Editors: D. P. Landau, K. K. Mon, and H.-B. Schüttler

76 *Computer Simulation Studies in Condensed-Matter Physics VI*
Editors: D. P. Landau, K. K. Mon, and H.-B. Schüttler

77 *Quantum Optics VI*
Editors: D. F. Walls and J. D. Harvey

78 *Computer Simulation Studies in Condensed-Matter Physics VII*
Editors: D. P. Landau, K. K. Mon, and H.-B. Schüttler

79 *Nonlinear Dynamics and Pattern Formation in Semiconductors and Devices*
Editor: F.-J. Niedernostheide

80 *Computer Simulation Studies in Condensed-Matter Physics VIII*
Editors: D. P. Landau, K. K. Mon, and H.-B. Schüttler

81 *Materials and Measurements in Molecular Electronics*
Editors: K. Kajimura and S. Kuroda

82 *Computer Simulation Studies in Condensed-Matter Physics IX*
Editors: D.P. Landau, K. K. Mon, and H.-B. Schüttler

Volumes 1–45 are listed at the end of the book

D. P. Landau K. K. Mon H.-B. Schüttler (Eds.)

Computer Simulation Studies in Condensed-Matter Physics IX

Proceedings of the Ninth Workshop
Athens, GA, USA, March 4–9, 1996

With 77 Figures

Springer

Professor David P. Landau, Ph. D.
Professor K. K. Mon, Ph. D.
Professor Heinz-Bernd Schüttler, Ph. D.

Center for Simulational Physics
The University of Georgia
Athens, GA 30602, USA

CIP data applied for

Die Deutsche Bibliothek - CIP-Einheitsaufnahme

Computer simulation studies in condensed matter physics : proceedings of the ... workshop ... - Berlin ; Heidelberg ; New York ; Barcelona ; Budapest ; Hong Kong ; London ; Milan ; Paris ; Santa Clara ; Singapore ; Tokyo : Springer.
9. Athens, GA, USA, March 4-9, 1996. - 1997
(Springer proceedings in physics ; 82)
NE: GT

ISBN-13: 978-3-642-64470-2 e-ISBN-13: 978-3-642-60597-0
DOI: 10.1007/978-3-642-60597-0

Typesetting: Camera ready copy from the authors/editors
SPIN: 10558039 54/3144 - 5 4 3 2 1 0 - Printed on acid-free paper

Preface

A decade ago, because of the phenomenal growth in the power of computer simulations, the University of Georgia formed the first institutional unit explicitly devoted to the use of simulations in research and teaching: the Center for Simulational Physics. We quickly sensed a need for a meeting place for both experienced simulators and neophytes to discuss new techniques and recent results in an environment that promoted extended discussion. As a consequence, the Center for Simulational Physics established an annual workshop on Recent Developments in Computer Simulation Studies in Condensed Matter Physics. This year's workshop was the ninth in this series, and the interest shown by the scientific community demonstrates quite clearly the useful purpose the series has served. The latest workshop was held at the University of Georgia, March 4–9, 1996, and these proceedings form a record which is published with the goal of timely dissemination of the material to a wider audience.

This volume is composed of three parts. The first section contains invited papers that deal with simulational studies of classical systems. The second section of the proceedings is devoted to invited papers on quantum systems, including new results for strongly correlated electron and quantum spin models. The final section comprises contributed presentations.

We hope that each reader will benefit from specialized results as well as profit from exposure to new algorithms, methods of analysis, and conceptual developments. We have already learned that fruitful collaborations and new resarch projects have resulted from these gatherings at the University of Georgia, and we hope that these proceedings will further expand this tendency.

Athens, GA, USA
June 1996

D.P. Landau
K.K. Mon
H.-B. Schüttler

Contents

Computer Simulation Studies in Condensed-Matter Physics: An Introduction

D.P. Landau, K.K. Mon, and H.-B. Schüttler

Center for Simulational Physics, The University of Georgia, Athens, GA 30602-2451, USA

The 9th Annual Workshop Proceedings are divided into three parts. The first two parts contain papers pertaining to invited presentations and include some pedagogical material. The final section contains short, contributed papers.

To open the first section of this volume, B. Dünweg, S. Kämmerer and M. Presber discuss recent work on the calculation of phase diagrams of models of binary alloys and adsorbed monolayers. Very large lattices of spins interacting with a nearest-neighbor Ising antiferromagnet Hamiltonian are used to study the Cu-Au system. Additional simulations use an elastic lattice gas model to study the unmixing phase transition of semiconductor alloys and for a two-dimensional model for c(2x2) structure formation. D. A. Huse uses a driven random-field XY model to study the physics of driven, sliding charge-density-wave systems or vortex lattices. This model permits topological defects and phase-slip, and it shows a sharp transition from regular, uniform sliding to chaotic, nonuniform, plastic sliding. K. Chen, T. Kaplan, and M. Mostoller review their recent large scale molecular dynamics simulations for two-dimensional Lennard-Jones systems on massively parallel computers. A recent reformulation of the NPT ensemble is used. In the largest systems of 36864 and 102400 atoms, a metastable hexatic phase was found between solid and liquid.

Several interesting new developments in computational studies of quantum systems have been presented in this workshop. The paper by N. Kawashima reports on the recent development of cluster flipping algorithms for quantum lattice systems. A specific application to the one-dimensional t-J model illustrates the substantial, order-of-magnitude reduction in autocorrelation times which can be achieved with these new methods. H. G. Evertz then reviews simulational results for various "quantum ladders," i.e., coupled quantum chains. These studies examine the dimensional crossover from one to two dimensions, and may shed light on theories for high T_c superconducting materials. S. Liang discusses a new method for treating correlated electron systems based on renormalization group (RG) ideas. He shows how an appropriate transformation of the single particle orbitals can generate a suitable hierarchy of energy scales in the problem which may then be exploited in an approximate RG approach. An application to an exactly solvable one-dimensional low-energy eigenstate calculations. In the final paper of this section, M. P. Nightingale reviews recent development of Monte Carlo methods in the calculation of eigenvalues for quantum Hamiltonians as well as transfer and stochastic matrices, starting from optimized trial eigenstates. Applications ranging from the groundstate energy calculation of quantum mechanical bosonic van der Waals clusters to the evaluation of static and dynamic critical exponents from thetransfer and stochastic matrices, respectively, of the Ising model are discussed.

A number of very interesting contributed papers follow. M. Kreer and P. Nielaba have studied the influence of substrate fields and quantum effects in graphite-adsorbed molecular hydrogen and deuterium, using path integral Monte Carlo methods. They present a detailed analysis of the anomalous transition temperature effects of the two systems which are experimentally observed in the commensurable $\sqrt{3}\times\sqrt{3}\,R30^\circ$ structure. The paper by M. H. Müser describes a path integral method for the simulation of quantum mechanical rotational

Springer Proceedings in Physics, Volume 82
Computer Simulation Studies in Condensed-Matter Physics IX
Eds.: D. P. Landau, K.K. Mon, H. -B. Schüttler

degrees of freedom, based on accurate numerical evaluations of the rotational kinetic energy time evolution kernel. Applications to a linear rotor model and to solid methane are discussed. T. Boutry, D. Ledue and D. P. Landau present a Monte Carlo study for the three-state Potts model on the two-dimensional quasi-periodic tiling. They report finite size-scaling results for the critical exponents which suggest that the model is in the universality class of the three-state Potts model on two-dimensional periodic lattices. K. Hukushima, H. Takayama and K. Nemoto describe an interesting new method, referred to as exchange Monte Carlo, for simulating poorly relaxing systems, exemplified here by the ±J Ising spin glass. The method is based on a simultaneous updating of many physically decoupled system replicas with each replica held at a different temperature. Compared to conventional updating schemes, their method appears to yield a dramatic speed-up in the Monte Carlo relaxation times. J. R. Ray and C. Frelėchoz describe a microcanonical ensemble theory with a rigorous statistical mechanics basis. As an example, the authors present results of simulations of a two-dimensional square lattice Ising model. I. Chang and H. Meirovitch use the scanning simulation method to study the θ transition for self-avoiding walks, terminally attached to an adsorbing linear boundary on a square lattice, with an attraction energy for non-bonded nearest-neighbor monomers. The simulations at both the special and ordinary points differ significantly from the exact values for the θ' model. K. Geisshirt, E. Praestgaard, and S. Toxvaerd outline how chemical reactions can be simulated by molecular dynamics. The technique is applied to a modified Volterra-Lotka scheme. Deviations from the phenomenological equations in the phase separating temperature region are observed. M. Laradji, S. Toxvaerd, and O. G. Mouritsen investigate the dynamics of phase separation in binary fluids. Molecular dynamics simulations are used to study the late time behavior and a growth exponent is estimated. B. L. Foley, A. M. Ferrenberg, and N. G. Adams use an event driven molecular dynamics method to simulate reacting plasmas. Techniques are developed which allow investigation of systems with up to 10^5 particles. G. A. Baker, Jr. and R. Gupta use extensive Monte Carlo calculations to examine finite size behavior in the three-dimensional Ising model near the critical point. In particular, an estimate for the renormalized coupling constant is extracted. J.-K. Kim applies a Wolff cluster-flipping Monte Carlo method to the two-dimensional XY-model. A novel finite size scaling analysis reveals inconsistencies in the simple description of the phase transition. The relaxation behavior of a small three-dimensional ±J spin glass is studied by T. Klotz. From the determination of the energy landscape the size of the transition matrix is dramatically reduced. S. Kobe then considers a method of combinatorial optimization for the search for the groundstate of a spin-glass. This "branch and bound" algorithm is successfully applied to small ±J spin glasses on a simple cubic lattice. M. Novotny uses a novel new MCAMC method (Monte Carlo with absorbing Markov chains) to study lifetimes of metastable states in the Ising square lattice. He shows that at low temperatures, discrete droplet theory, modified by a temperature independent prefactor, holds.

Part I

Classical Systems

Phase Diagrams of Alloys and Adsorbed Monolayers: Some Recent Results

B. Dünweg, S. Kämmerer, and M. Presber

Institut für Physik, Universität Mainz, Postfach 3980,
D-55099 Mainz, Germany

We discuss some recent work done on the calculation of phase diagrams of models of binary alloys and adsorbed monolayers. For the nearest–neighbor Ising antiferromagnet on the fcc lattice (model for the Cu–Au system) we study a rather large lattice of 4×64^3 spins. This is necessary since the inherent frustration of the lattice induces a very small interfacial tension between ordered domains. We find no indications for the suggested L' phase, and locate the triple point at a *nonzero* temperature. There is some numerical evidence that it might in fact be a *multicritical* point. We then discuss the extension of lattice gas models to "elastic lattice gases" (ELGs) which include also translational degrees of freedom. Special attention is paid to the statistical treatment of vacant sites, and it is shown that a system A + vacancies is no longer equivalent to a system A + B. The ELG Hamiltonian is then studied for three–dimensional models on the diamond lattice for the unmixing of semiconductor alloys (where we find Mean–Field–like critical behavior), and for a two–dimensional model for $c(2 \times 2)$ structure formation.

1 Introduction: Lattice Gas and Ising Models

Although originally invented for magnetic systems, the Ising model is, in many cases, not a particularly good representation for real magnets, since its spin dimensionality $n = 1$ is too low. Conversely, it is very well suited to describe substitutional binary alloys (AB) or adsorbed monolayers, since in both cases there is a "pre–defined" lattice given by either the mixed crystal (three–dimensional alloy) or the adsorption sites of the substrate surface (two–dimensional adsorbate system), and there are two states per lattice site available (A–atom vs. B–atom in the alloy case, adsorbate atom vs. vacancy in the monolayer case). This analogy is made rigorous by the lattice gas model with pair interactions,

$$\begin{aligned} \mathcal{H}_{LG} &= -\sum_i \left\{ \varepsilon^A c_i + \varepsilon^B (1-c_i) \right\} \\ &+ \sum_{\langle ij \rangle} \left\{ v_{nn}^{AA} c_i c_j + v_{nn}^{AB} \left[c_i(1-c_j) + c_j(1-c_i) \right] + v_{nn}^{BB} (1-c_i)(1-c_j) \right\} \\ &+ \sum_{\langle\langle ij \rangle\rangle} \left\{ v_{nnn}^{AA} c_i c_j + v_{nnn}^{AB} \left[c_i(1-c_j) + c_j(1-c_i) \right] + v_{nnn}^{BB} (1-c_i)(1-c_j) \right\} + \ldots \end{aligned} \qquad (1)$$

Here, $c_i = 1$ if site i is occupied by an A–atom, while $c_i = 0$ otherwise (in the adsorbate case, B is just a vacancy). $\langle ij \rangle$ and $\langle\langle ij \rangle\rangle$ denote pairs of nearest and of next–nearest neighbors, respectively, and of course the model could also include interactions of even larger range. ε^A is the energy which is released upon placing an A–atom on a site, while v_{nn}^{AA} is the energy needed to build a nearest neighbor bond of two A–atoms. The other symbols' meaning is analogous. This Hamiltonian is quite general, and for an alloy one would specify $\varepsilon^A = \varepsilon^B = 0$, while for an adsorbate $\varepsilon^B = v_{\ldots}^{AB} = v_{\ldots}^{BB} = 0$.

Springer Proceedings in Physics, Volume 82
Computer Simulation Studies in Condensed-Matter Physics IX
Eds.: D. P. Landau, K.K. Mon, H. -B. Schüttler

Now this model is straightforwardly mapped onto an Ising model [1] by introducing the pseudospin variables $S_i = 2c_i - 1 = \pm 1$, and studying the model of N sites in the grand–canonical ensemble, whose partition function is

$$Z_{gc} = \sum_{N_A=0}^{N} \exp(\beta\mu_A N_A)\exp(\beta\mu_B N_B) Z_c(N_A), \tag{2}$$

where β is the inverse temperature $1/(k_B T)$, μ_A and μ_B denote the chemical potentials of species A and B, respectively (of course, $\mu_B = 0$ in the adsorbate case), and

$$Z_c(N_A) = \sum_{\{c_i\}|N_A} \exp\left(-\beta\mathcal{H}_{LG}\right) \tag{3}$$

is the canonical partition function, the sum running over all configurations compatible with the constraint $\sum_i c_i = N_A$ (which automatically means $\sum_i(1-c_i) = N_B = N - N_A$, such that in the alloy case one should use, strictly spoken, the term "semi–grand canonical ensemble"). With this transformation one finds, apart from an irrelevant prefactor

$$Z_{gc} = \sum_{\{S_i\}} \exp\left(-\beta\mathcal{H}_I\right) \tag{4}$$

with

$$\mathcal{H}_I = -J_{nn}\sum_{\langle ij\rangle} S_i S_j - J_{nnn}\sum_{\langle\langle ij\rangle\rangle} S_i S_j - \ldots - H\sum_i S_i \tag{5}$$

and

$$\begin{aligned} H &= \frac{1}{2}(\mu_A - \mu_B) + \frac{1}{2}\left(\varepsilon^A - \varepsilon^B\right) \\ &\quad -\frac{z_{nn}}{4}\left(v_{nn}^{AA} - v_{nn}^{BB}\right) - \frac{z_{nnn}}{4}\left(v_{nnn}^{AA} - v_{nnn}^{BB}\right) - \ldots, \end{aligned} \tag{6}$$

$$J_{nn} = \frac{1}{4}\left(2v_{nn}^{AB} - v_{nn}^{AA} - v_{nn}^{BB}\right), \tag{7}$$

$$J_{nnn} = \frac{1}{4}\left(2v_{nnn}^{AB} - v_{nnn}^{AA} - v_{nnn}^{BB}\right), \tag{8}$$

etc. Here z_{nn} and z_{nnn} denote the coordination numbers of the lattice in the nearest and next–nearest neighbor shell, respectively.

Though its apparent simplicity, $\mathcal{H}_I$ allows for a very rich physics. This is so because the model allows for competing interactions and resulting complicated antiferromagnetic structures. For a given set of J's, the usual Monte Carlo analysis of such a Hamiltonian proceeds as follows: (i) Find the ground states as a fuction of H; (ii) identify the phases, the pertinent order parameters (usually linear combinations of sublattice magnetizations), and their symmetry properties; (iii) from that, conclude the universality class of second–order transitions, if applicable (e. g. Ising, XY with cubic anisotropy, etc.); (iv) calculate the phase diagram. This program looks rather straightforward, and in many cases it is. However, there are cases where the intrinsic complexity of $\mathcal{H}_I$ causes quite challenging problems, and one such case will be discussed in some detail in the next Section.

Nevertheless, despite these rather interesting properties, the above approach is clearly limited. One obvious deficiency is the symmetry of the resulting phase diagram: From the Ising symmetry (invariance of $\mathcal{H}_I$ with respect to $S_i \to -S_i$ and $H \to -H$), one concludes that the phase diagram in the grand–canonical ensemble (H–T plane) is symmetric around $H = 0$, while the corresponding canonical phase diagram (c–T plane, $0 \le c \le 1$ being the concentration of A–atoms) must be symmetric around $c = 0.5$. However, the phase diagrams of real systems are often strongly asymmetric, and the traditional approach to remedy this "sickness" has been to include also triplet interactions, i. e. terms $\propto S_i S_j S_k$.

However, this symmetry will also be destroyed as soon as the model also includes *elastic* interactions, i. e. explicitly takes into account the translational degrees of freedom of each particle. Moreover, this modification also changes the *range of the effective interaction* between the spins *qualitatively* because two spins, although being located rather far away from each other, nevertheless interact with each other due to the long–range elastic distortions of the lattice. For this reason, these systems are also of fundamental physical interest, since the universal critical behavior of second–order phase transitions may well be affected. The rest of this contribution will therefore discuss some recent studies of such models. Sec. 3 will be devoted to the modeling and the related methodological problems, while Sec. 4 and 5 discuss the application to a three–dimensional model for a mixture of Si and Ge (note that the effects of elastic interactions, i. e. most notably the atomic size mismatch, are expected to be most pronounced in covalently bonded systems like semiconductor alloys), and to a two–dimensional model for an adsorbed monolayer of hydrogen on a palladium (100) surface, respectively.

2 The Nearest–Neighbor Ising Antiferromagnet on the FCC Lattice

We study the Hamiltonian of Eqn. 5 on the fcc lattice for the case $J = J_{nn} < 0$, while all other exchange couplings vanish, and we limit ourselves to $H > 0$ for symmetry reasons. A physical realization of this system is the alloy Cu_xAu_{1-x}, i. e. in this system the same ordered superstructures occur [2]. The phase diagram of this model has been the subject of a long–standing debate, and calculations have been done, with ever–increasing accuracy, using the Mean Field approximation or Kikuchi's cluster variation method (CVM) [3–9], low–temperature expansions [10–12] and Monte Carlo (MC) simulations [13–20, 9]. The present MC study [21, 22] is the most expensive simulation of the system so far, and has attempted to resolve the controversial issues about the location of the triple point and about the existence of the L' phase (see below).

The reason why this system has been so hard to analyze is the geometric *frustration* of the fcc lattice: It is impossible to assign spins to a nearest–neighbor tetrahedron such that all six bonds are antiferromagnetic. Therefore, the ground state [23] does not exhibit three–dimensional order but only two–dimensional order (except for $H > 12|J|$, where all spins are up), i. e. it is a sequence of perfectly ordered (either ferromagnetic or antiferromagnetic) (100) planes. Every antiferromagnetic plane yields a twofold degeneracy (since it may be shifted freely in itself), such that the ground state is macroscopically degenerate with nevertheless vanishing entropy. For $H < 4|J|$, all planes are antiferromagnetic, while for $4|J| < H < 12|J|$ every second plane is ferromagnetic. At the "superdegenerate" point $H = 4|J|$ the ground state entropy is nonzero.

For $T > 0$, three–dimensional order becomes *entropically* stabilized [10–12], since the ordered phases AB ($H < 4|J|$) and A_3B ($4|J| < H < 12|J|$) admit more low–energy excitations than a disordered ground state. As usual, a configuration of two ordered domains separated by two interfaces (also called antiphase boundaries, APBs; note the periodic boundary conditions) costs a free energy penalty of $\Delta F = 2\sigma L^2$, where L is the linear system size and σ the interfacial tension [24]. However, the pathology is that σ has a purely entropic origin, and hence vanishes quickly upon $T \to 0$. For this reason, L must be rather large in order to make ΔF sufficiently large, and practical tests [21] showed that we needed an $N = 4 \times L^3$ system with $L = 64$, while $L = 32$ still showed some tendency towards APBs, and $L = 16$, which had been used in older studies [13–15], is clearly too small.

In order to describe the ordered phases, we decompose the system into four interpenetrating simple cubic sublattices (a, b, c, d), and introduce

$$\begin{aligned}\psi_0 &= (m_a + m_b + m_c + m_d)/4 \\ \psi_1 &= (m_a + m_b - m_c - m_d)/4 \\ \psi_2 &= (m_a - m_b + m_c - m_d)/4 \\ \psi_3 &= (-m_a + m_b + m_c - m_d)/4,\end{aligned} \tag{9}$$

where m_α are the sublattice magnetizations varying between -1 and $+1$. ψ_0 is the total magnetization, and ψ_1, ψ_2 and ψ_3 are the components of the three–dimensional order parameter $\vec{\psi}$, which vanishes in the disordered phase where all sublattices are equivalent. The perfectly ordered AB state is then given by two sublattices with $S_i = +1$, the other two sublattices having magnetization -1. Likewise, in the A_3B state only one sublattice has magnetization -1, all the other spins being $S_i = +1$. The AB phase is hence described by the six states $\vec{\psi} = (\pm\psi_{AB}, 0, 0)$, $(0, \pm\psi_{AB}, 0)$ and $(0, 0, \pm\psi_{AB})$, and the A_3B phase by the four states $\vec{\psi} = (\psi_{A3B}, \psi_{A3B}, \psi_{A3B})$, $(\psi_{A3B}, -\psi_{A3B}, -\psi_{A3B})$, $(-\psi_{A3B}, \psi_{A3B}, -\psi_{A3B})$, $(-\psi_{A3B}, -\psi_{A3B}, \psi_{A3B})$, where $|\psi_{AB}| < 1$ and $|\psi_{A3B}| < 1/2$ describe states which are not perfectly ordered. The so–called L' phase corresponds to *three* different sublattice magnetizations (ideally, two sublattices up, one down, and one with random spin orientation); this phase has been found stable by CVM calculations in the vicinity of $H = 4|J|$ [7]. We carefully searched for this phase at various state points by starting the system in a perfect L' state, and monitoring the time evolution of the sublattice magnetizations. In all cases, the system ended up in an AB or A_3B state, and hence we believe that this phase is an artifact of the CVM.

Similarly, the old MC studies [13–15] had suggested that the triple point, where AB, A_3B and the disordered phase coexist, should occur at $H = 4|J|$, $T = 0$. We hence looked, in a similar fashion, also for stability of the disordered phase near $H = 4|J|$ at low temperatures, with the same result, i. e. evolution into an AB or A_3B state. Therefore, the triple point must occur at a *nonzero* temperature. We then mapped out the phase boundaries, using standard hysteresis loops and thermodynamic integration to find intersecting branches of the free energy F, using the relations $N\psi_0 = -\partial F/\partial H$ and $U = -T^2\partial(F/T)/\partial T$ for the internal energy.

Note that all transition lines are expected to be of first order: The six states of the AB phase correspond to a Heisenberg model with cubic anisotropy, whose transition into the disordered phase has been shown to be first–order by renormalization–group arguments [25]. Similarly, the A_3B phase corresponds to a 4–state Potts model (relevant for the transition into the disordered phase), while the transition from A_3B to AB corresponds to a 3–state Potts model (note that one out of three sublattices must be flipped) [26].

Our results are summarized in Fig. 1. One sees that the data indeed confirm the anticipated first–order behavior of all three transition lines, and the triple point is located at $k_BT_t/|J| = 0.98 \pm 0.02$, $H_t/|J| = 3.60 \pm 0.04$. Moreover, we find that all first–order jumps along all three lines get very small when approaching the triple point. We hence believe that this point might actually be a *multicritical* point, such that the jumps would tend to *zero*. Indeed, our order parameter data along the AB $\leftrightarrow$ A_3B transition line are well compatible with tricritical scaling. However, our resolution is not fine enough to unambiguously answer this subtle question; this would require further large–scale simulations of even significantly larger lattices. We believe that an accurate finite–size scaling analysis would prove very difficult, due to the high order parameter dimensionality, and the very large minimum system size which is necessary. It should also be noted that a multicritical point is *not* predicted by Landau theory [22].

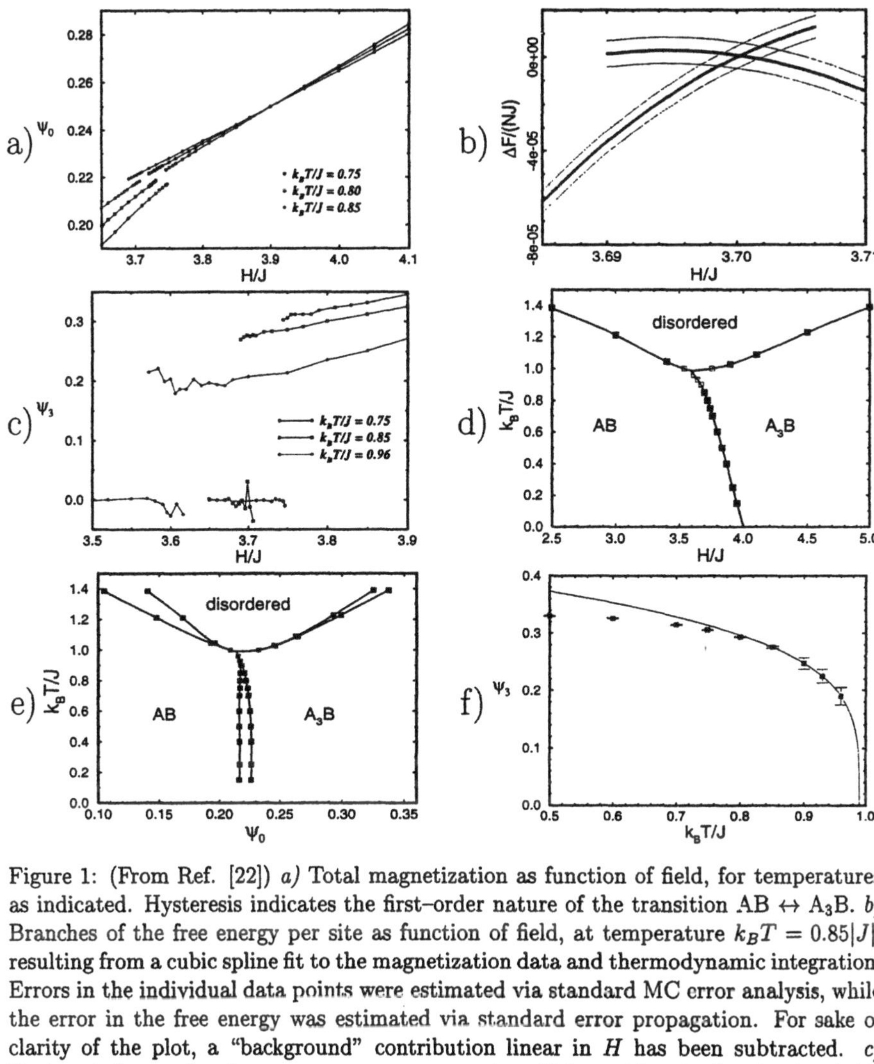

Figure 1: (From Ref. [22]) *a)* Total magnetization as function of field, for temperatures as indicated. Hysteresis indicates the first–order nature of the transition AB $\leftrightarrow$ A_3B. *b)* Branches of the free energy per site as function of field, at temperature $k_BT = 0.85|J|$, resulting from a cubic spline fit to the magnetization data and thermodynamic integration. Errors in the individual data points were estimated via standard MC error analysis, while the error in the free energy was estimated via standard error propagation. For sake of clarity of the plot, a "background" contribution linear in H has been subtracted. *c)* Hysteresis loops of order parameter component ψ_3, which vanishes in the simulated AB state, while it is nonzero in the A_3B phase. The first–order jumps decrease strongly upon approaching the triple point. *d)* Phase diagram in the grand–canonical ensemble. Filled symbols have been obtained by thermodynamic integration, open symbols by direct inspection of order parameter hystereses. *e)* Phase diagram in the canonical ensemble. *f)* Order parameter component ψ_3, evaluated along the AB $\leftrightarrow$ A_3B coexistence line, as a function of temperature. The line represents a fit $\psi_3 \propto (T_t - T)^{\beta_t}$ with $k_BT_t = 0.99|J|$ and $\beta_t = 0.2416$, while tricritical scaling would imply $\beta_t = 1/4$. However, every exponent in the range $0.2 \ldots 0.3$ is compatible with the data.

3 Inclusion of Elastic Interactions

The "fundamental" approach to a classical simulation of an alloy or adsorbate system with continuous degrees of freedom would consist in the specification of an interatomic potential (depending on distances, bond angles, atomic species, ...) and, for modeling the influence of a substrate, an external potential. The particles would then freely move around, and, depending on the current geometric configuration, find the neighbors with which they interact. One could also introduce a grand–canonical ensemble via particle creation and destruction, or a semi–grand canonical description via changes A $\leftrightarrow$ B. Indeed, for fluids, this approach *must* be used, and has recently been applied to rather accurate studies of the gas–liquid transition of two– and three–dimensional Lennard–Jones systems [27, 28].

However, it is also obvious that such a model is computationally rather cumbersome: Each particle's neighbor shell fluctuates and must be continuously checked and updated. We have hence chosen to deliberately neglect these fluctuations and instead study a two– or three–dimensional *network* with the same *topology* as the original lattice. Since this topology is conserved during the simulation, one can use the same neighbor list for the interactions once and for all, such that from the point of view of data structures the simulation is as efficient as a simple lattice simulation. Actually, the types of Hamiltonians we are interested in are directly related to the original lattice gas, Eqn. 1: One simply has to replace the interaction parameters v_{nn}^{AA} etc. by distance–depending functions,

$$v_{nn}^{AA} \to v_{nn}^{AA}\left(|\vec{r}_i - \vec{r}_j|\right) \tag{10}$$

(similarly the other v's), and the adsorption energies ε^A, ε^B by functions depending on the particle's distance from its ideal adsorption site:

$$\varepsilon^A \to \varepsilon^A\left(\left|\vec{r}_i - \vec{r}_i^{\,id}\right|\right). \tag{11}$$

Again, for describing an alloy one would specify $\varepsilon^A = \varepsilon^B = 0$, while in the adsorbate case $\varepsilon^B = v_{...}^{AB} = v_{...}^{BB} = 0$. The resulting Hamiltonian shall be denoted with $\mathcal{H}_{ELG}$ ("elastic lattice gas").

Of course, such a simplified Hamiltonian loses some of the physics which could otherwise be modeled by the translational degrees of freedom, since both structural phase transitions as well as melting are explicitly excluded. However, sufficiently far away from such transitions we expect the model to describe the physics reasonably well.

This model has a number of peculiarities which shall now be discussed. First of all, one should note that the particles should be considered as *distinguishable*: Each particle can be uniquely identified by its position on the network, and hence prefactors $(N_A!)^{-1}$, $(N_B!)^{-1}$ do *not* occur in the partition function. Moreover, the model's "philosophy" is to assign *independent* degrees of freedom c_i and $\vec{r}_i$ to each lattice site i, and to also update these degrees of freedom independently of each other in the grand–canonical ensemble. However, if the B–particle is a vacancy, this assignment is unphysical and artificial, since physically it does not make sense to ask for the coordinate of a non–existent particle. Nevertheless, this assignment is useful in order to facilitate a compact and straightforward simulation algorithm. This is the reason why the transformation to the grand–canonical ensemble is no longer as straightforward as before, and why the systems A + B (alloy) and A + vacancies (adsorbate) *are no longer equivalent* [29].

Let us now discuss the transformation to the grand–canonical ensemble in some more detail. Eqn. 2 is of course still valid, but Z_c must now also include the translational degrees of freedom of the *real* particles. To make this more explicit, consider the lattice sites as enumerated from 1 to N. A given configuration $\{c_i\}$ can then be associated in a unique way with indices $i_1 < i_2 < \ldots < i_{N_A}$ and $j_1 < j_2 < \ldots < j_{N_B}$ such that $c_{i_k} = 1$

and $c_{j_k} = 0$. Moreover, we introduce the numbers $\chi_{nn}(i,j) = 1$ if i and j are nearest neighbors, and zero otherwise, and similarly $\chi_{nnn}(i,j)$ for next–nearest neighbors, etc. $\mathcal{H}_{ELG}$ can then be rewritten as

$$\begin{aligned}
\mathcal{H}_{ELG} &= -\sum_{k=1}^{N_A} \varepsilon^A\left(\left|\vec{r}_{i_k} - \vec{r}_{i_k}^{id}\right|\right) - \sum_{k=1}^{N_B} \varepsilon^B\left(\left|\vec{r}_{j_k} - \vec{r}_{j_k}^{id}\right|\right) \\
&+ \sum_{1\leq k<l\leq N_A} \chi_{nn}(i_k,i_l) v_{nn}^{AA}\left(\left|\vec{r}_{i_k} - \vec{r}_{i_l}\right|\right) \\
&+ \sum_{k=1}^{N_A}\sum_{l=1}^{N_B} \chi_{nn}(i_k,j_l) v_{nn}^{AB}\left(\left|\vec{r}_{i_k} - \vec{r}_{j_l}\right|\right) \\
&+ \sum_{1\leq k<l\leq N_B} \chi_{nn}(j_k,j_l) v_{nn}^{BB}\left(\left|\vec{r}_{j_k} - \vec{r}_{j_l}\right|\right) \\
&+ \ldots,
\end{aligned} \tag{12}$$

where the next–nearest neighbor and higher contributions are written in the same way as the nearest neighbor terms. In the alloy case we then find for the canonical partition function, using an arbitrary normalization volume V_0,

$$\begin{aligned}
Z_c^{alloy}(N_A) &= \sum_{\{c_i\}|N_A} V_0^{-N_A} V_0^{-N_B} \int d\vec{r}_{i_1} \ldots \int d\vec{r}_{i_{N_A}} \times \\
&\quad \times \int d\vec{r}_{j_1} \ldots \int d\vec{r}_{j_{N_B}} \exp\left(-\beta\mathcal{H}_{ELG}\left(\{\vec{r}_{i_k}\}, \{\vec{r}_{j_l}\}\right)\right) \\
&= \sum_{\{c_i\}|N_A} V_0^{-N} \int d\vec{r}_1 \ldots \int d\vec{r}_N \exp\left(-\beta\mathcal{H}_{ELG}\left(\{c_i\}, \{\vec{r}_i\}\right)\right),
\end{aligned} \tag{13}$$

such that, apart from a constant prefactor,

$$Z_{gc}^{alloy} = \sum_{\{c_i\}} V_0^{-N} \int d\vec{r}_1 \ldots \int d\vec{r}_N \exp\left(-\beta\mathcal{H}_{eff}\left(\{c_i\}, \{\vec{r}_i\}\right)\right) \tag{14}$$

with

$$\mathcal{H}_{eff} = \mathcal{H}_{ELG} - (\mu_A - \mu_B)\sum_i c_i. \tag{15}$$

I. e. in this case the same transformation as in the simple lattice gas case applies. $\mathcal{H}_{eff}$ can, of course, be used directly to control a standard Metropolis algorithm. In the adsorbate case with vacancies, however, only the coordinates of the *real* particles appear:

$$Z_c^{adsorb}(N_A) = \sum_{\{c_i\}|N_A} V_0^{-N_A} \int d\vec{r}_{i_1} \ldots \int d\vec{r}_{i_{N_A}} \exp\left(-\beta\mathcal{H}_{ELG}\left(\{\vec{r}_{i_k}\}\right)\right). \tag{16}$$

We now construct an effective Hamiltonian $\mathcal{H}_{eff}$ by *requiring* that the grand–canonical partition function Z_{gc}^{adsorb} can also be written in the form of Eqn. 14. Then $\mathcal{H}_{eff}$ can be used directly for a standard Metropolis scheme, too. We start with the *ansatz* [29]

$$\mathcal{H}_{eff} = \mathcal{H}_{ELG} - (\mu_A + \alpha k_B T)\sum_i c_i + \sum_i (1 - c_i) U_0\left(\left|\vec{r}_i - \vec{r}_i^{id}\right|\right). \tag{17}$$

The physical motivation is that without the counterterms $\alpha > 0$, $U_0 \geq 0$ particle destruction processes become more and more likely with increasing temperature and increasing system size, for entropical reasons: The ghost particles can explore the full system volume, while the real particles are confined to their immediate neighborhood. The purpose of U_0 is to confine also the ghost particles around their adsorption sites, while the term

α provides for the necessary temperature dependence (for higher temperatures, the creation of real particles becomes more and more favored by this term). Inserting Eqn. 17 into Eqn. 14, one finds, after splitting the integration into real coordinates $\vec{r}_{i_k}$ and ghost particle coordinates $\vec{r}_{j_l}$,

$$Z_{gc}^{adsorb} = \sum_{N_A=0}^{N} \exp\left(\beta\mu_A N_A\right) \exp\left(N_A \alpha + N_B \ln \zeta\right) Z_c^{adsorb}(N_A), \tag{18}$$

where we have introduced the single-particle partition function resulting from the potential U_0,

$$\zeta = V_0^{-1} \int d\vec{r} \exp\left(-\beta U_0(\vec{r})\right). \tag{19}$$

Obviously, a correct simulation algorithm is obtained for $\alpha = \ln \zeta$ (note $\mu_B = 0$). We have tested [29] two choices of U_0, (i) $U_0 = 0$, i. e. $\alpha = \ln(V/V_0)$, where V is the total system volume, and (ii) a square-well potential which confines the particle to a volume V_c, i. e. $\alpha = \ln(V_c/V_0)$. It turned out that only the second choice is useful, for purely *dynamic* reasons: If no confining potential is used, the ghost particles are free to travel through all of the system. However, as soon as they are some lattice constants away from their ideal site, a rematerialization would place them onto a very high energy level, due to the strong potential ε^A (for which we used a harmonic spring). Hence, this rematerialization is forbidden by the Boltzmann factor, and does not occur until the random walker happens to come close to its ideal site again. The correlation time of the algorithm should be roughly proportional to the time needed for these "loops", and this should, from diffusion arguments, scale like L^2 if L is the linear system size. I. e. this algorithm has intrinsic critical slowing down built in, independent of the state point! The square-well potential obviously removes this problem, while, per construction, it nevertheless yields correct static averages. Note also that the arbitrary normalization parameter V_0 enters the simulation algorithm. This means that the chemical potential μ_A is only defined after V_0 has been specified, i. e. a change of V_0 corresponds to a temperature-dependent redefinition of the zero of μ_A. Hence, for such simulations the μ_A-T phase diagram needs the additional specification of V_0.

One also sees rather easily that the Ising symmetry will in general no longer hold for the ELG: After transforming to pseudospin variables, the Hamiltonian assumes the form

$$\mathcal{H} = \mathcal{H}_0\left(\{\vec{r}_i\}\right) - \sum_{\langle ij \rangle} J_{ij}\left(\{\vec{r}_i\}\right) S_i S_j - \sum_{\langle\langle ij \rangle\rangle} J_{ij}\left(\{\vec{r}_i\}\right) S_i S_j - \ldots - \sum_i H_i\left(\{\vec{r}_i\}\right) S_i, \tag{20}$$

and it is impossible to induce a change $H_i \to -H_i$ via just changing the chemical potential (note also that for the so-called "compressible Ising model" [30] $H_i \equiv 0$). Hence, we do not view the Ising language as particularly useful for the ELG. For adsorbed monolayers, the "elastic" route towards phase diagram asymmetries has already qualitatively been noted by Persson [31].

4 The Si–Ge Unmixing Transition

For a binary alloy of silicon and germanium, we studied an ELG on the diamond lattice. Apart from nearest-neighbor pair interactions, we also introduced three-body interactions. This was *not* done in order to induce a phase diagram asymmetry, but rather in order to stabilize the fourfold coordinated structure of the diamond lattice, i. e. the tetrahedron angle. The first simulation [32] used the Keating potential [33]:

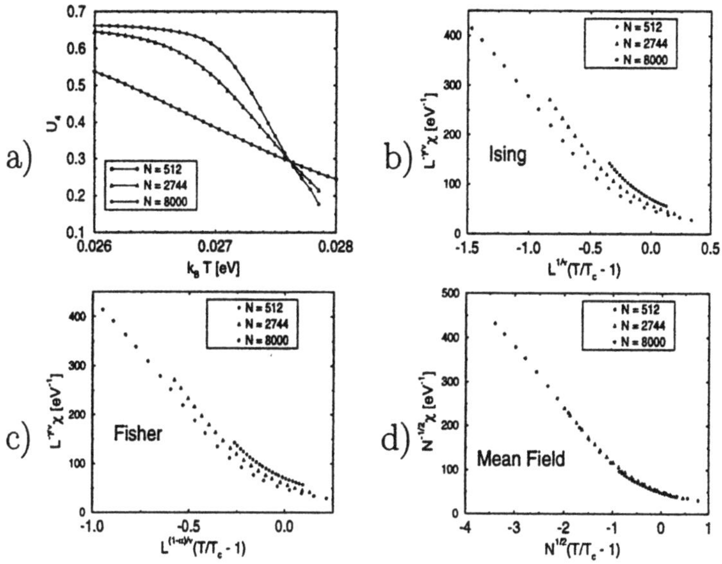

Figure 2: (From Ref. [32]) Critical behavior of the ELG modeling a Si–Ge alloy using the Keating potential: *a)* Maximum value of the cumulant as a function of temperature, for three system sizes as indicated. *b)* Data collapsing plot for the susceptibility, checking for Ising–like critical behavior. *c)* Same as b) for Fisher–renormalized exponents. *d)* Same as b) for Mean Field critical behavior.

$$v_{nn}^{\alpha\beta}(r) = \phi_{nn}^{\alpha\beta} + E^{\alpha\beta}\left[r^2 - \left(l_0^{\alpha\beta}\right)^2\right]^2 \tag{21}$$

with $\alpha, \beta \in \{A, B\}$. The corresponding three–body term is

$$v_3^{\alpha\beta\gamma} = A^{\alpha\beta\gamma}\left[\vec{r}_{ij} \cdot \vec{r}_{kj} + \frac{1}{3} l_0^{\alpha\beta} l_0^{\gamma\beta}\right]^2, \tag{22}$$

i. e. the summation runs over all angles with vertex at site j, occupied with species β (note $\cos\Theta_t = 1/3$ for the tetrahedron angle Θ_t). For the interaction parameters, see Ref. [32]. This model was simulated at constant vanishing pressure [34] in order to accomodate the atomic size mismatch, for systems of up to $N = 8 \times 10^3$ sites. The unmixing phase diagram was calculated using thermodynamic integration procedures [32].

The critical behavior of the model was analyzed by a multi–histogram analysis [35] of data taken near the critical point. The results are shown in Fig. 2. First, we calculated the fourth–order cumulant of the (extensive) magnetization (in Ising language),

$$U_4 = 1 - \frac{\left\langle (M - \langle M \rangle)^4 \right\rangle}{3\left\langle (M - \langle M \rangle)^2 \right\rangle^2}, \tag{23}$$

as a function of both chemical potential and temperature and determined, at fixed temperature, its maximum, the location defining the critical chemical potential. Plotting the resulting values as a function of temperature yields Fig. 2 a), showing that the standard

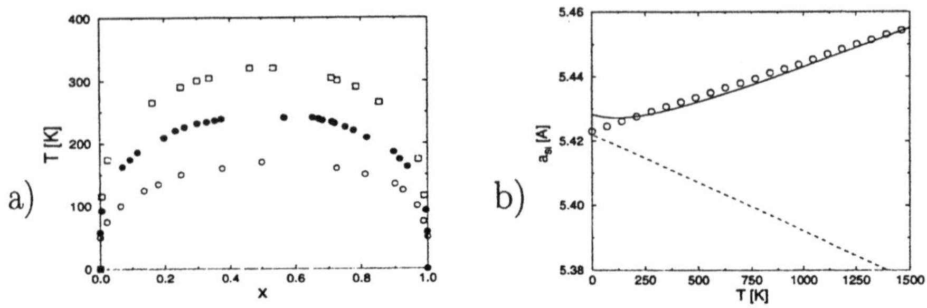

Figure 3: (From Ref. [39]) *a)* Unmixing phase diagram of $Si_{1-x}Ge_x$ alloys using different models: Keating potential [32] (open squares), SW potential [39] (solid circles), Tersoff potential [41]. This latter potential is even significantly more complicated, and the critical behavior had not been studied in this paper. *b)* Lattice constant of pure Si as a function of temperature, for Keating potential (dashed line), SW potential (open circles), and fit to experiments (solid line).

cumulant intersection method for an accurate determination of the critical point works also for the ELG.

The finite–size scaling relation for the susceptibility reads in the Ising case

$$\chi = (Nk_BT)^{-1}\left(\left\langle M^2\right\rangle - \langle M\rangle^2\right) \tag{24}$$

$$\chi = L^{\gamma/\nu}\tilde{\chi}\left(tL^{1/\nu}\right), \tag{25}$$

with $t = T/T_c - 1$, $\gamma = 1.24$, $\nu = 0.63$, while for Fisher–renormalized exponents [36] (which are not expected in the present case) one has to replace $\gamma \to \gamma_{eff} = \gamma/(1-\alpha)$, $\nu \to \nu_{eff} = \nu/(1-\alpha)$ with $\alpha = 0.12$. In the Mean Field case one has [37]

$$\chi = L^{d/2}\tilde{\chi}\left(tL^{d/2}\right). \tag{26}$$

As seen from Fig. 2 b)–d), the data are in best agreement with Mean Field critical behavior. This is probably due to an effective long–range interaction induced by the elastic distortions. It is expected that this corresponds to a dipolar interaction, for which the upper critical dimension is $d = 3$ [38].

It should be mentioned that a similar study [39] was also done for the Stillinger–Weber (SW) potential [40]. The same Mean–Field–like behavior was found; however, the SW potential yields a different critical temperature, and a more physical behavior of thermal expansion, see Fig. 3. For more details, see Refs. [32, 39].

5 Hydrogen on Pd(100)

We used a simple two–dimensional ELG (adsorbate and vacancies) on the square lattice to model, e. g., the adsorption of H on Pd, taking into account both nearest as well as next–nearest neighbor interactions [29]. The Hamiltonian is specified via simple harmonic springs,

$$\varepsilon^A(r) = \frac{k_0}{2}r^2 \tag{27}$$

$$v_{nn}^{AA}(r) = \phi_{nn} + \frac{k_{nn}}{2}(r - l_{nn})^2 \tag{28}$$

and

$$v_{nnn}^{AA}(r) = \phi_{nnn} + \frac{k_{nnn}}{2}(r - l_{nnn})^2 \tag{29}$$

where the lattice constant is set to unity, and the springs' rest lengths are, for simplicity, assumed to be fully compatible with the substrate lattice, $l_{nn} = 1$ and $l_{nnn} = \sqrt{2}$. We choose $\phi_{nn} = +4$ and $\phi_{nnn} = -4$ in dimensionless units, and, for simplicity, $k_0 = k_{nn} = k_{nnn} = 1$. The ghost particles were confined to circular areas of unit radius around the ideal adsorption sites (cf. Eqn. 17). Likewise, we set the normalization volume V_0, which defines the chemical potential for $T > 0$, to $V_0 = 1$.

For $k_i = 0$ the model reduces to a LG whose properties had been studied a long time ago [42]. The phase diagram's topology (see Fig. 4) is unchanged by the elastic effects. There occurs just one ordered structure, $c(2 \times 2)$, i. e. a nearest–neighbor antiferromagnet (cf. Fig. 5), whose transition to the disordered phase is of second order for high temperatures, while it is of first order for low temperatures. Interestingly enough, the overall

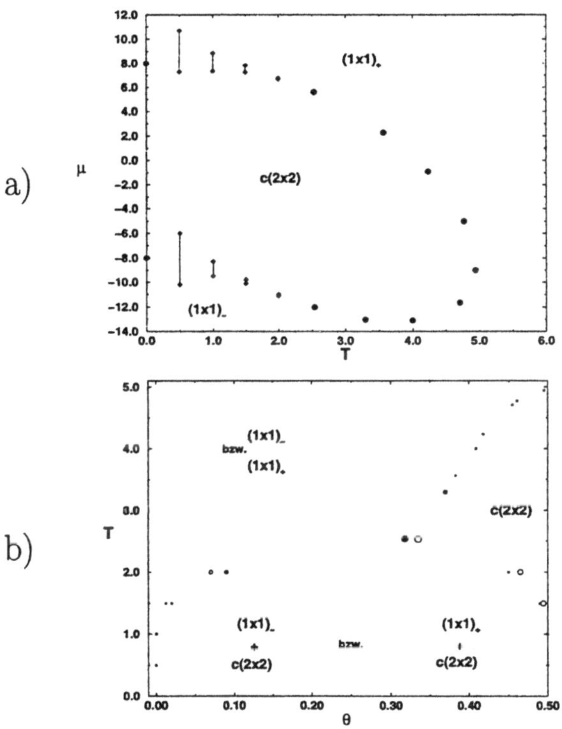

Figure 4: (From Ref. [29]) *a)* Phase diagram of the two–dimensional model specified in Eqns. 27–29, in the grand–canonical ensemble. The asymmetry is not surprising, in view of the chemical potential for $T > 0$ being unique only up the normalization volume V_0 (cf. Sec. 3). Solid circles denote second–order transitions, while diamonds denote hysteresis ranges of first–order transitions. *b)* Phase diagram of the same model in the canonical ensemble; $0 \leq \Theta \leq 1$ is the coverage. Solid circles, $0 \leq \Theta \leq 0.5$; open circles, $0.5 \leq \Theta \leq 1$. Data for Θ and $1 - \Theta$ have been included in the same figure in order to make the weak asymmetry more visible. Errors are always of the order of symbol sizes.

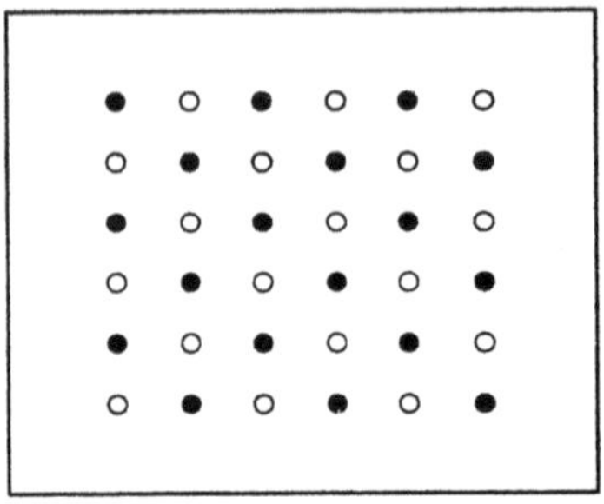

Figure 5: (From Ref. [29])
The $c(2 \times 2)$ structure.

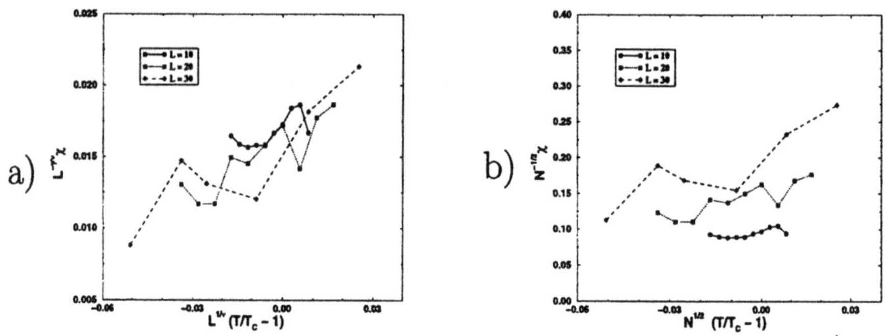

Figure 6: (From Ref. [29]) *a)* Data collapsing plot of the staggered susceptibility for 2d Ising exponents, at $T_c = 4.231$. *b)* Same as a) for Mean field behavior.

asymmetry of the canonical phase diagram is rather weak. Moreover, the overall shape is only weakly changed in comparison with the pure LG [42].

The second–order phase transitions were analyzed by standard finite–size scaling of the susceptibility of the staggered magnetization, i. e. the difference in sublattice magnetizations of the two sublattices separated by nearest–neighbor bonds, using system sizes $L = 10, 20, 30$. The data are not accurate enough to draw firm conclusions, but it seems that they are in better agreement with Ising–like behavior than with Mean Field, see Fig. 6. This would further corroborate the hypothesis that the upper critical dimension is $d = 3$ for the ELG.

Acknowledgements

The work reported here was done in close collaboration with D. P. Landau, K. Binder, M. Laradji, and M. d'Onorio de Meo. B. D. also gratefully acknowledges support from the Alexander von Humboldt foundation, and from a NATO travel grant.

References

[1] K. Binder, in *Adv. Sol. State Physics*, edited by P. Grosse (Vieweg, Braunschweig, 1986), Vol. 26, pp. 133–168.

[2] F. Ducastelle, *Order and Phase Stability in Alloys* (North Holland, Amsterdam, 1991).

[3] W. Shockley, J. Chem. Phys. **6**, 130 (1938).

[4] U. Gahn, Z. Metallkunde **64**, 268 (1973).

[5] R. Kikuchi, J. Chem. Phys. **60**, 1071 (1974).

[6] J. M. Sanchez, D. de Fontaine, and W. Teitler, Phys. Rev. B **26**, 1465 (1982).

[7] A. Finel and F. Ducastelle, Europhys. Lett. **1**, 135 (erratum 543) (1986).

[8] A. Finel, in *Alloy Phase Stability*, edited by G. M. Stocks and A. Gonis (Kluwer, Dordrecht, 1989), pp. 269–280.

[9] R. Tetot, A. Finel, and F. Ducastelle, J. Stat. Phys. **61**, 121 (1990).

[10] J. Slawny, J. Stat. Phys. **20**, 711 (1979).

[11] J. Slawny, in *Phase Transitions and Critical Phenomena*, edited by C. Domb and J. L. Lebowitz (Academic Press, London, 1987), Vol. 11, pp. 127–205.

[12] J. Bricmont and J. Slawny, J. Stat. Phys. **54**, 89 (1989).

[13] K. Binder, Phys. Rev. Lett. **45**, 811 (1980).

[14] K. Binder, Z. Phys. B **45**, 61 (1981).

[15] K. Binder, J. L. Lebowitz, M. K. Phani, and M. H. Kalos, Acta Metall. **29**, 1655 (1981).

[16] U. Gahn, J. Phys. Chem. Solids **43**, 977 (1982).

[17] U. Gahn, J. Phys. Chem. Solids **47**, 1153 (1986).

[18] H. Ackermann, S. Crusius, and G. Inden, Acta Metall. **34**, 2311 (1986).

[19] H. T. Diep, A. Ghazali, B. Benge, and P. Lallemand, Europhys. Lett. **2**, 603 (1986).

[20] J. L. Lebowitz, M. K. Phani, and D. F. Styer, J. Stat. Phys. **38**, 413 (1985).

[21] S. Kämmerer, Diplom thesis, Johannes Gutenberg–Universität Mainz, 1994.

[22] S. Kämmerer, B. Dünweg, K. Binder, and M. d'Onorio de Meo, Phys. Rev. B **53**, 2345 (1996).

[23] O. J. Heilmann, J. Phys. A **13**, 263 (1980).

[24] K. Binder, Z. Phys. **B 43**, 119 (1981).

[25] D. J. Wallace, J. Phys. C **6**, 1390 (1973).

[26] E. Domany, Y. Shnidman, and D. Mukamel, J. Phys. C **15**, L495 (1982).

[27] N. B. Wilding and A. D. Bruce, J. Phys.: Condens. Matter **4**, 3087 (1992).

[28] N. B. Wilding, Phys. Rev. E **52**, 602 (1995).

[29] M. Presber, Diplom thesis, Johannes Gutenberg-Universität Mainz, 1996.

[30] D. J. Bergman and B. I. Halperin, Phys. Rev. B **13**, 2145 (1976).

[31] B. N. J. Persson, Surf. Sci. **258**, 451 (1991).

[32] B. Dünweg and D. P. Landau, Phys. Rev. B **48**, 14182 (1993).

[33] P. N. Keating, Phys. Rev. **145**, 637 (1966).

[34] W. W. Wood, in *Physics of Simple Liquids*, edited by H. N. V. Temperley, J. S. Rowlinson, and G. S. Rushbrooke (North Holland, Amsterdam, 1968).

[35] A. M. Ferrenberg and R. H. Swendsen, Phys. Rev. Lett. **63**, 1195 (1989).

[36] M. E. Fisher, Phys. Rev. **176**, 257 (1968).

[37] E. Brezin and J. Zinn-Justin, Nucl. Physics **B 257**, 867 (1985).

[38] A. Aharony, in *Phase Transitions and Critical Phenomena*, edited by C. Domb and M. S. Green (Academic Press, London / New York / San Francisco, 1976), Vol. 6.

[39] M. Laradji, D. P. Landau, and B. Dünweg, Phys. Rev. B **51**, 4894 (1995).

[40] F. H. Stillinger and T. A. Weber, Phys. Rev. B **31**, 5262 (1985).

[41] P. C. Kelires and J. Tersoff, Phys. Rev. Lett. **63**, 1164 (1989).

[42] K. Binder and D. P. Landau, Surf. Sci. **108**, 503 (1981).

Plasticity and Chaos in Models of Randomly-Pinned, Driven Lattices

D.A. Huse

Department of Physics, Princeton University,
Princeton, NJ 08544, USA

Abstract. A relatively simple model that captures some of the physics of driven, sliding charge-density-wave systems or vortex lattices is a driven random-field XY model. This model permits topological defects and phase-slip. Some of the previous results on these systems are reviewed and then simulation results are presented for a three-dimensional model. The model shows a fairly sharp transition from regular, uniform sliding with little phase-slip for weaker pinning to chaotic, nonuniform, plastic sliding with much more phase-slip at stronger pinning.

1. Introduction

There are number of solid-state systems where two types of crystallization are present at low temperatures and the two crystals can be driven to slide past one another. The first level of crystallization is simply the atomic or molecular crystallinity of a given material. We will here be particularly interested in the realistic case where this first level of crystallization is not perfect, there being static chemical and/or structural imperfections in the material. The second level of "crystallization" is that of an order parameter, such as the electronic charge density in a material that exhibits a charge-density-wave (CDW) ordering [1,2] or Wigner crystallization of its electrons, or the density of vortices in the Abrikosov vortex-lattice phase of a type-II superconductor [3]. An electric field can drive the motion of a charge-density-wave or a Wigner crystal, while an electrical transport current can drive the motion of the vortex lattice in a type-II superconductor.

The case considered here is where the spatial period of the order parameter pattern (the second level of crystallization) is incommensurate with the lattice spacings of the underlying atomic crystal. In this case the system's free energy is independent of the relative displacement of the two crystals when they are both free of static defects or imperfections. Such perfect, incommensurate crystals will exhibit a smooth, continuous relative "sliding" motion in linear response to a driving force, as long as the interactions between the two crystals are weak enough. However, any defects or imperfections that are immobile in either lattice will generally break this degeneracy, producing an energy that does depend on the relative displacements of the two lattices. This will inhibit or stop the sliding motion, "pinning" the relative positions of the lattices. This pinning also induces distortions in the lattices and, when it is strong enough, can induce defects. In this paper I will review and investigate the behavior of these systems as the strength of the pinning and the driving force are varied.

Springer Proceedings in Physics, Volume 82
Computer Simulation Studies in Condensed-Matter Physics IX
Eds.: D. P. Landau, K.K. Mon, H. -B. Schüttler

Generally in these systems the elastic constants of the underlying atomic crystal are much larger than those of the order-parameter pattern, so it is a good approximation (and a convenient simplification) to assume the atomic crystal is perfectly static and rigid. Also, one is generally interested in low temperatures, far below the melting transition of the material, where any motion of the chemical or structural defects in the atomic lattice is extremely slow. Thus, again, it is both a good approximation and a convenient simplification to assume that these defects are perfectly stationary. Both of these assumptions are adopted here. The simplest case to consider is when the ideal, unpinned order-parameter pattern is nonuniform (and periodic) along only one spatial direction, as in a uniaxial charge-density-wave (CDW). Most of the discussion here will, for simplicity, be in terms of such a uniaxial CDW system, driven by a uniform electric field, E. It is only the motion along the direction of the periodicity of the CDW that represents motion of the CDW. E is the component of the electric field that drives this motion. In this simple case, the degrees of freedom of the system are the local amplitude and displacement of the CDW.

To be a little more precise, let $\rho(\mathbf{r})$ be the local charge density. An ideal, unpinned, uniaxial CDW is $\rho(\mathbf{r}) = \rho_0 + A_0 cos(\mathbf{k_0} \cdot \mathbf{r})$, where $\mathbf{k_0}$ is a wavevector. (More generally, the cosine function is replaced by a periodic function that may contain higher harmonics.) A general configuration that is not too strongly deformed from this ideal pattern can be written as $\rho(\mathbf{r}) = \rho_0 + A(\mathbf{r}) cos(\mathbf{k_0} \cdot \mathbf{r} + \phi(\mathbf{r}))$, and we may remove the rapid variation with wavevector $\mathbf{k_0}$, defining the order-parameter pattern as $\psi(\mathbf{r}) = A(\mathbf{r}) exp(i\phi(\mathbf{r}))$. In the absence of pinning the local energy and dynamics will be independent of a spatially uniform translation of the CDW, which corresponds to a rotation of the order parameter ψ in the complex plane. Thus we have a single continuous symmetry as in an XY model. However, the pinning breaks this continuous symmetry, locally favoring a particular displacement of the CDW and thus a particular value of the phase ϕ. The random pinning is a random symmetry-breaking field, like a random local magnetic field in the XY model. The uniform electric field, E, drives the CDW to move, and thus drives the order parameter ψ to rotate in the complex plane. The simplest type of model that includes all these effects is a driven random-field XY model.

Let us consider a CDW configuration that is represented by a smooth (on some microscopic length scale) configuration of the order parameter, $\psi(\mathbf{r})$. In general, the phase, $\phi(\mathbf{r})$, is only defined modulo 2π, and is not well-defined at locations where ψ vanishes. In a three-dimensional system, the manifolds along which ψ vanishes (if they exist) are generically lines. These lines are vortices in the phase configuration: the phase increases or decreases continuously by 2π as one encircles a vortex line. Away from any vortex lines the phase is smoothly varying, so its gradient, $\nabla\phi(\mathbf{r})$, is well-defined. Let us define the phase difference between two points, ($\mathbf{r}$, $\mathbf{r}'$), as the line integral of the phase gradient along the straight line between the two points. This phase difference will discontinuously increase or decrease by 2π whenever a vortex line crosses the straight line between the two points. This process is often called "phase-slip" [4,5,2].

What types of behavior can a CDW system or model exhibit? 1) The CDW can be stationary, as occurs if the pinning is strong enough or the drive is weak enough, and thermal and quantum fluctuations can be ignored. Such a stationary configuration may or may not contain vortices. 2) The CDW may be sliding with the local, time-averaged sliding velocity being spatially uniform and nonzero. The local sliding velocity is proportional to the rate of phase change, $d\phi(\mathbf{r})/dt$. Such a uniform sliding motion can in principle continue indefinitely without building up large phase differences, even in the absence of phase-slip processes. Lastly, 3) The CDW may be sliding with the local, time-averaged sliding velocity being spatially nonuniform (including possibly some regions that are stopped). For this nonuniform sliding to continue indefinitely, phase-slip and thus vortex motion must be occuring to prevent the build-up of arbitrarily large phase differences between points with different local sliding velocities. This last regime of nonuniform or plastic motion may be further divided into subregimes based on whether or not a nonzero fraction of the system is sliding together at precisely the same nonzero time-averaged velocity and whether or not the dynamical system is chaotic, as is discussed below.

The remainder of this paper contains a brief review of some of the previous work on this problem and then a report of simulation results on a simple driven XY or CDW model, focussing particularly on the regime of chaotic, nonuniform, plastic sliding motion with phase-slip and on the transition between this chaotic regime and the regime of more regular and uniform sliding.

2. Statics

A first question to consider is what are the properties of the CDW when it is stationary, in particular when it is not being driven. This general problem was first addressed by Larkin [6] in the context of pinned vortex lattices, then by Imry and Ma [7] for random-field magnets and by Fukuyama, Lee and Rice [4] for CDWs. A simple model to examine is the random-field XY model, with energy

$$H = -h\sum_i \cos(\beta_i - \phi_i) - \sum_{<ij>} \cos(\phi_i - \phi_j), \tag{1}$$

where i and j are sites of a regular lattice, ϕ_i is the local phase of the CDW at site i, the local pinning at site i prefers $\phi_i = \beta_i$, the pinning strength is h, and the second sum runs over all nearest-neighbor pairs $< ij >$ on the lattice. Any variation between lattice sites of the amplitude of the CDW is ignored in this simple model, so only the phase degree of freedom remains. The pinning phases, β_i, are assumed to be random and static, independently and uniformly distributed modulo 2π. To define vortices in this model, let us use the convention that the phase difference $(\phi_i - \phi_j)$ between nearest-neighbor sites always is in the interval $(-\pi, \pi)$. Then add these phase differences along the nearest-neighbor bonds of any elementary loop (plaquette) of the lattice. This sum is an integer multiple of 2π, and when it is nonzero this integer is the net vorticity enclosed by that plaquette (a vortex counts as +1 and an antivortex counts as −1).

In the absence of the random pinning, $h = 0$, any state of spatially uniform phase is a stable minimum of the energy (1). For weak pinning, $h << 1$, let us first consider states that are smoothly and gradually deformed away from such a uniform state, so the phase differences between nearest-neighbor sites are small. A stable state is one that is a local minimum of the energy, H, so the force, $-dH/d\phi_i$, at each site vanishes. Certainly, if one starts in a stable state and advances all the phases by 2π one is again in a stable state, but this is really just the same state as before, because the phases are only defined modulo 2π. However, one may instead advance the phase by roughly 2π over only a region of linear size L, to return it to the locally stable state, and let this phase change gradually drop to zero over a length scale of order L as one moves away from this region. Can such a state also be stable? The total net force over this region of size L from the second term in (1), that wants to eliminate this phase deformation, is of order L^{d-2} for a d-dimensional system. The total pinning force from the first term in (1) is a sum of L^d random forces of order h, so can be of order $hL^{d/2}$. For small h and L the pinning force is not enough to stabilize such a deformation, but for the physical case $d < 4$, there is a pinning length, varying as $L_p \sim h^{-2/(4-d)}$ for small h, beyond which the pinning forces can balance the "elastic" restoring force and stabilize the new deformed state. Such stable deformations should be possible almost everywhere in the system, producing many distinct stable states for small h.

Among the stable states of (1) there is one that has the lowest energy, the ground state. For zero pinning, $h = 0$, the ground state is degenerate and is any state of spatially uniform phase. For strong pinning, $h >> 1$, the ground state will have each phase ϕ_i nearly aligned with the local pinning phase β_i. This latter state has high density of vortices, while the ground state for $h = 0$ has no vortices. How does the ground state evolve between these two limits? The nature of the pinned ground state with vortices forbidden has been recently studied [8,9]. It has been suggested that for small, nonzero pinning h the ground state for a d = 3-dimensional system remains vortex-free at large scales even when vortices are permitted [9–11]. If this is correct, then there is a topological phase transition at a nonzero critical pinning, $h_c > 0$, where arbitrarily long vortex loops first appear in the ground state of (1) on an infinite three-dimensional lattice. Monte Carlo simulations of the random-field XY model are consistent with this scenario [11]. This proposed ground-state transition is the static counterpart of the transition in the sliding state from a uniformly sliding state, without large-scale phase-slip due to moving vortices, to a nonuniformly sliding state, which must have moving vortices to produce the necessary phase-slip. Discussion and simulations of this regime in the sliding state follow below.

3. Dynamics

A simple model for a sliding CDW is to give the above random-field XY model (1) relaxational dynamics and a uniform driving field, E, at zero temperature:

$$d\phi_i/dt = E - dH/d\phi_i. \tag{2}$$

A further simplification is to replace $cos(\phi_i - \phi_j)$ in the nearest-neighbor interaction in (1) with its quadratic expansion, $1-(\phi_i-\phi_j)^2/2$. This eliminates the 2π periodicity of the energy with varying ϕ_i, and thereby forbids phase-slip and vortices. Such models without vortices or phase-slip permitted have a sharp depinning transition separating the pinned and uniformly sliding states [12,13]. However, Coppersmith and Millis [14,15] demonstrated that in such models arbitrarily large nearest-neighbor phase differences appear near the depinning transition and in the sliding state. Thus the approximation of truncating the expansion of the cosine at quadratic order necessarily breaks down and phase-slip processes occur, destroying the sharp depinning transition and introducing the regime of nonuniform sliding motion.

The nonuniform sliding motion arises because of the nonuniformity of the random pinning [14,15]. A given local random pinning pattern (pattern of the β_i's here) if repeated over the whole system will yield some sliding rate, call it $\omega_{local}(\mathbf{r})$. For small drive and strong enough pinning there is no possible local pinning configuration that would allow sliding, so $\omega_{local} = 0$ everywhere and there are (one or more) stationary states. These stationary states are fixed points of the dynamical system defined by (2). However, when the drive is larger or the pinning weaker, the probability distribution of ω_{local} will no longer be a δ-function. Generally the smallest possible ω_{local} comes from the strongest pinning configuration, which is uniform pinning, i.e., all β_i's equal. The weakest possible pinning is likely some regular, perhaps quasiperiodic [15], pattern of β_i's and produces the largest ω_{local}. Following Coppersmith and Millis' argument [15], let us assume the CDW is uniformly sliding with rate Ω. Consider a region of volume L^d with $\omega_{local} \neq \Omega$. This region on its own will slide at rate ω_{local} so the remainder of the system must exert a net force of $(\Omega - \omega_{local})L^d$ on it to keep it sliding at rate Ω. This force must be carried by the nearest-neighbor couplings across the region's surface; the required force per bond is of order $(\Omega - \omega_{local})L$. But the maximum force such a nearest-neighbor bond in (1) can support is unity. Whenever ω_{local} has a probability distribution with nonzero width, there is a nonzero (although perhaps extremely small) probability of a region existing that is large enough so that the required force per surface bond exceeds one. The necessary force then can not be delivered to keep such a region sliding at rate Ω, and it instead slides at a different rate, producing phase-slip.

The argument in the preceding paragraph applies to systems where the sliding does not involve the transport of a conserved quantity [15]. If there is a conserved quantity that is necessarily transported if and only if there is sliding, then there can be additional forces, corresponding to a pressure conjugate to the density of the conserved quantity, that can serve to hold the anomolously-pinned regions against phase-slip. In a CDW system the electrons can move coherently as part of the CDW and they may also move incoherently as "normal" carriers. Thus the conservation of charge does not strictly constrain the CDW motion: the CDW in a region may slide coherently, and the resulting current be converted to and from a normal (non-CDW) current at the boundary of the sliding region. For a sliding vortex lattice, the net vorticity does appear to be a conserved quantity, so the simple model we are considering here, that ignores such conservation laws, may miss effects due to this constraint.

Although the above argument has ruled out uniform sliding for our model (2), the regions that are sliding at different rates may be very rare and dilute. These rare regions may not be strong enough to disrupt the remainder of the system from sliding together at precisely the same rate. Following on previous numerical [16,17] and experimental [18] work, Koshelev and Vinokur [19] suggested that sliding vortex lattices may have a dynamical phase transition from a state of nonuniform, plastic sliding to a uniformly sliding vortex crystal as the driving force and sliding speed are increased. For the CDW model these two sliding phases may be differentiated based on the presence or absence of a sort of temporal long-range order, as argued by Balents and Fisher [20]. For each site we may define its time-averaged sliding rate, $\omega_i =< d\phi_i/dt >$, and also the local power spectrum, $S_i(\omega) = |\hat{\psi}_i(\omega)|^2$, where $\hat{\psi}_i(\omega)$ is the time fourier transform of $\psi_i(t) = exp(i\phi_i(t))$. In the sliding state with temporal long-range order, the average of this local power spectrum over all sites, $[S_i(\omega)]$ has a "narrow-band", δ-function component at a nonzero global sliding frequency Ω. Balents and Fisher [20] argue that such a temporally-ordered sliding phase can exist with random pinning for $d \geq 3$-dimensional systems. A related, but not necessarily equivalent criterion for temporal order is whether a finite fraction of the sites have precisely the same nonzero time-averaged local sliding rate, ω_i. This latter question is investigated in the simulations reported below.

The mean-field version of (2) with every site equally coupled to every other site has been investigated by Strogatz, *et al.* [21]. In this mean-field limit there is no nonuniform motion due to unusually pinned regions, because there are no regions. Each site is just coupled to its local pinning and the mean field. This mean-field model shows three regimes of behavior: a pinned phase, a uniformly sliding phase, and a bistable regime where both the pinned and sliding states are dynamically stable [21].

4. Simulation results

The model I have numerically studied is

$$d\phi_i/dt = E + hsin(\beta_i - \phi_i) + \sum_j^{(i)} \sin(\phi_j - \phi_i), \qquad (3)$$

where i is any site on the simple cubic lattice ($d = 3$) and the sum is over its six nearest neighbors, j. The external drive pushing the phase ϕ_i at all sites to advance is uniform and of strength E. The local pinning strength is h at all sites and tends to pin ϕ_i at the angle β_i; these pinning angles, β_i, are quenched, random, and uniformly distributed modulo 2π. This model is approximated simply at first-order in time when it is integrated on the computer, so the system actually examined is explicitly

$$\phi_i(t + \delta t) = \phi_i(t) + \delta t\{E + hsin(\beta_i - \phi_i(t)) + \sum_j^{(i)} \sin(\phi_j(t) - \phi_i(t))\}. \qquad (4)$$

Periodic boundary conditions are used, with cube-shaped lattices of L^3 sites, with $L \leq 45$. Quantities measured include the time-averaged sliding rate at each site, $\omega_i =< d\phi_i/dt >$, its mean, variance and spatial correlation functions; the largest Lyapunov exponent, λ_{max}, of this dynamical system; and the density

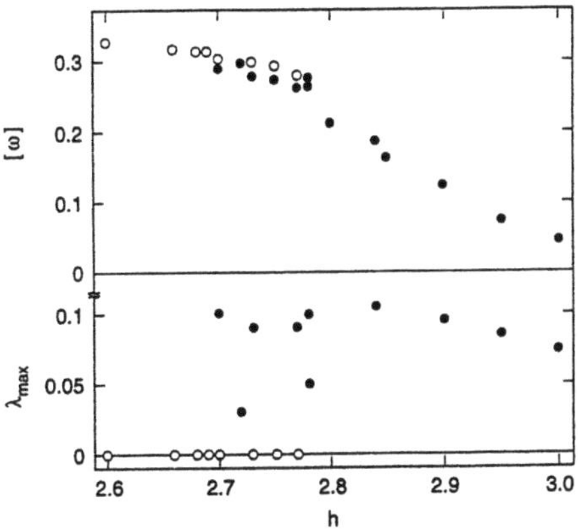

Figure 1. The sliding rate, $[\omega]$, averaged over time and all sites, and the largest Lyapunov exponent, λ_{max}, vs. the pinning strength h, at drive $E = 0.5$. The open circles represent nonchaotic states, while the filled circles represent chaotic states. For each value of h where there are both types of state shown, these two represent the same realization of the random pinning, but with different initial conditions. However, different realizations of the random pinning are used at each different value of h.

of vortices in the configuration of ϕ. The initial conditions are either uniform, $\phi_i(0) = \epsilon$, or aligned with the random pinning, $\phi_i(0) = \epsilon + \beta_i$. The system is then advanced in time for $\epsilon = 0$, calculating the trajectory, $\phi_i(t)$, and its derivative with respect to a small uniform perturbation in the initial condition, $d\phi_i(t)/d\epsilon$. The magnitude of this dependence on the initial condition (rms over all i) varies as $exp(\lambda_{max} t)$ at long times.

The model (4) has three parameters: $\delta t, E$ and h. δt should be chosen small enough to give a reasonable approximation to (3); in practice I have used $\delta t = 0.1$. I have surveyed various values of the other parameters, settling on a more detailed study for varying pinning strength h at fixed drive $E = 0.5$. All the behaviors seen elsewhere seem to be represented along this cut through parameter space at fixed drive. Note that increasing the pinning at fixed drive gives qualitatively the same sequence of behavior as decreasing the drive at fixed (and large enough) pinning.

Results for the regime $2.6 \le h \le 3.0$, where most of the changes in behavior are observed, are shown in Fig. 1. For weak pinning, $h \le 2.5$, the system is attracted to a simple limit cycle with all sites sliding at the same rate, no phase-slip, and $\lambda_{max} = 0$. Although the above argument for nonuniform motion should apply for $h \neq 0$ in this regime, the regions sliding at a different rate are presumably so rare that they did not occur in the samples of size $45^3 \cong 10^5$ sites that were examined. On increasing the pinning, h, slightly nonuniform sliding was first seen at $h = 2.6$, where less than 0.1% of the sites were observed to be stopped ($\omega_i = 0$), with the remainder all sliding at the same rate. This still constitutes a limit cycle behavior with $\lambda_{max} = 0$. The fraction of stopped sites increases steadily with increasing $h > 2.6$; see Fig. 2.

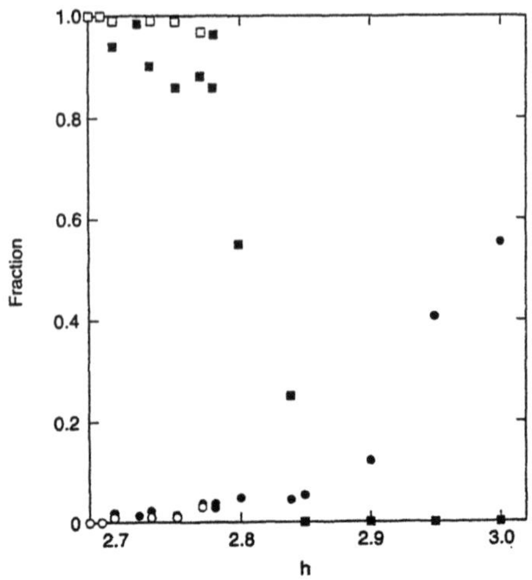

Figure 2. The fraction of sites stopped (circles) and the fraction of sites sliding together at the same rate (squares) vs. pinning strength h. As in Fig. 1, the open symbols represent nonchaotic states, the filled symbols represent chaotic states, and the drive is $E = 0.5$.

At $h = 2.70$ two new types of behavior were observed that were not seen for $h \leq 2.69$. One sample of 45^3 sites, started from the randomly pinned initial condition, remained in a chaotic ($\lambda_{max} \cong 0.10$) sliding state for the entire run which lasted over 20,000 time units. In this chaotic state the distribution of sliding rates appears to consist of two δ-functions with a continuous part between them. Roughly 2% of the sites were stopped, while roughly 94% were sliding at the same rate within a part in 10^3. When this same sample was instead started from a uniform initial condition, it went to a nonchaotic state, sliding slightly faster, with $\lambda_{max} = 0$, and fewer (about 1.5%) of the sites stopped. In this nonchaotic state, a few (in this case, 3) of the sites were found to be sliding at nonzero rates less than that adopted by the vast majority (about 98.5%) of the sites. This apparent bistability between chaotic and nonchaotic states was also observed for some, but not all, other samples of the same size (45^3) in the range $2.70 \leq h \leq 2.77$. For $h \geq 2.78$ the nonchaotic sliding state was no longer found to be stable, even after starting from uniform initial conditions. As h is increased beyond this range of apparent bistability, the average sliding rate, $[\omega]$, falls off fairly rapidly (Fig. 1). The fraction of the sites that are sliding together at the same rate falls off quite abruptly (Fig. 2), apparently vanishing by $h = 2.85$. The number of stopped sites grows steadily with h (Fig. 2), exceeding half the sites by $h = 3.0$ and exceeding 90% of the sites by $h = 3.2$.

The difference in the average sliding rate between the chaotic and nonchaotic states in the regime of bistability is actually rather small compared to the changes in the sliding rate over the h interval shown in Fig. 1. This is also true for the vortex density and for the distribution of the local sliding rates. On the

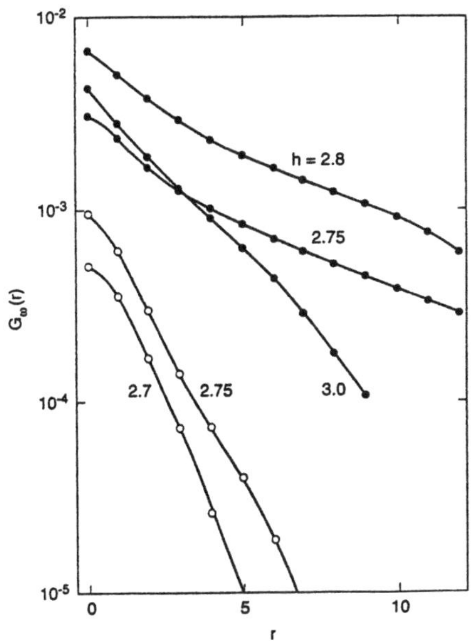

Figure 3. The correlation function of the local time-averaged sliding rate, $G_\omega(r)$, defined by Eq. (5), for various values of the pinning strength, h. As in Figs. 1 and 2, the open symbols represent nonchaotic states, the filled symbols represent chaotic states, and the drive is $E = 0.5$. For pinning strength $h = 2.75$ the chaotic and nonchaotic results are for the same realization of the random pinning, but different initial conditions. The curves in this figure only serve to connect the points that represent the same sliding state.

other hand, for some samples the difference in the largest Lyapunov exponent, λ_{max}, across the bistable regime is almost as large as its full variation with h; see Fig. 1. The other measured quantity that shows a relatively large difference between the "coexisting" states in the bistable regime is the correlation function of the local sliding rates,

$$G_\omega(r) = [\omega_i \omega_j] - [\omega_i]^2, \tag{5}$$

where the average in the first term is over all pairs of sites, (i, j), that are separated by distance r lattice units. As shown in Fig. 3, in the nonchaotic states the correlation length is well under 2 lattice units, while in the coexisting chaotic states it is at least 4 lattice units. This is the largest the correlation length gets; the long-distance correlations reach a maximum near $h = 2.8$ and then decline with further increase in h.

Although a careful study of the sample-size dependence of these results has not yet been done, examination of some smaller samples shows that the results are substantially size- and sample-dependent only in the interval $2.7 \leq h \leq 2.9$. The general tendency is for smaller samples to be sliding more uniformly, faster and less chaotically. This allows the nonchaotic states to survive to somewhat higher h for the smaller samples. An important ingredient in the chaos is presumably having different regions sliding with different frequencies, which then interact nonlinearly, producing chaos. Smaller samples will have fewer different sliding frequencies represented, so are less likely to be chaotic.

Some of the results reported here are possibly not representative of a true steady-state, they could instead be very long-lived transients. This is most likely to be a concern for the chaotic states in and just above the bistable regime. A few runs did show long-lived chaotic transients, the system eventually either falling onto a nonchaotic attractor at long times or moving to a state with a different sliding speed and degree of chaos and plasticity. The nonchaotic states are periodic (or quasiperiodic in some cases), do not show strong fluctuations and have $\lambda_{max} = 0$ to numerical (single-) precision, so certainly appear dynamically stable. Well above the transition regime, for $h > 2.9$ there do not appear to be any long transients or substantial finite-size effects, so the states accessed in this study are very likely a good representation of the true steady-state of an infinite system in this stronger-pinning regime.

It is interesting to speculate about what happens in the infinite-system and infinite-time limit. One important question is when, if ever, do a finite fraction of the sites slide at precisely the same rate? Well into the stronger-pinning, plastic regime, for $h \geq 2.85$, the fraction found to slide together at the same rate is only a few per cent, and is dropping steadily with time as the sliding rates are measured more precisely by averaging over more cycles. Thus it appears quite certain that there is no temporal long-range order in this stronger-pinning regime.

In the weak-pinning regime, on the other hand, temporal long-range order should be present for the $d = 3$-dimensional system [20]. The nonuniformity of the pinning will induce strains in the sliding state. The tail of the distribution of strains represents the rare regions where phase-slip occurs [14,15]. However, at weak enough pinning these regions are quite rare and for $d > 2$ the typical local strain is much too small to induce phase-slip. Thus it appears that in the weaker-pinning regime the bulk of the system is sliding at the same rate, with temporal long-range order, and there is only a low density of isolated regions that are sliding at different rates.

Where is the phase transition between temporally ordered and disordered states located? One possibility is that these two types of states can actually both be dynamically stable at the same drive and pinning strength in the infinite system, and the bistable regime observed here for finite systems reflects that. However, as shown in Fig. 2, in the bistable regime 80% or more of the sites were found to be sliding at the same rate to within one part in 10^3 in the chaotic states, suggesting that these chaotic states have temporal order. Thus it appears that the temporal long-range order may survive past the observed bistable regime and into the fully chaotic regime. A more thorough numerical study is needed to adequately answer this question, perhaps using some form of finite-size scaling and measurements of other quantities than those examined here.

One can generalize the Coppersmith-Millis [14,15] argument for phase-slip to argue that chaos and multistability will also always be present in the sliding state. When the regions with phase-slip occur with nonzero density, they can, with nonzero probability, be near enough to each other to interact strongly enough to produce chaos and multistability. These chaotic regions are thus a subset of the regions with phase-slip, so when the phase-slip is rare, the chaos is even more rare. Thus in the present study of a few finite samples there is a regime in the temporally ordered phase, namely $2.6 \leq h \leq 2.69$, where phase-slip is common enough to have occurred, but chaos is still so rare that it was not seen. As h is increased further, phase-slip and chaos become more common, and chaos is seen in the present study, although it appears to remain too dilute to destroy the temporal long-range order until h reaches 2.8 or so. This argument suggests that the onset of chaos and bistability, like the onset of phase-slip, is due to rare regions and does not represent the phase transition where the temporal long-range order is lost. The results shown in Fig. 2 are consistent with this picture: the strong drop in the fraction of sites sliding at the same rate occurs at somewhat larger values of the pinning strength h than the bistable regime and the onset of chaos.

In conclusion, I have numerically studied the zero-temperature, three-dimensional driven random-field XY model (4) as a simple CDW model that permits phase-slip. For weak enough pinning or strong enough drive, the system shows a nonchaotic sliding motion with all but a very small minority of the sites sliding at the same time-averaged sliding rate. This constitutes a temporal long-range order [20]. For stronger pinning, the sliding state is chaotic with a continuous distribution of local sliding rates, much phase-slip and plasticity, and no temporal long-range order. In the vicinity of the transition between these two types of sliding states, some of the samples studied showed bistability, where both chaotic and nonchaotic sliding states appeared to be dynamically stable for the same values of the pinning strength and drive. The onset of the chaos and bistability appears to occur in the phase with temporal long-range order, where a nonzero fraction of the sites are sliding at the same rate. Thus there appears to be a narrow intermediate regime with both chaos and temporal order. The sliding state in this regime, if it does indeed exist, presumably consists of an infinite connected cluster of sites sliding at the same rate with no phase-slip, surrounding regions of phase-slip and chaos that are sufficiently dilute and/or weak that they do not break up this infinite cluster. The transition that destroys the temporal order is then when the density and/or strength of the phase-slip and chaos become large enough to destroy the infinite cluster. This description suggests that this transition might be viewed as a sort of percolation transition. However, a more thorough understanding of this type of phase transition will require further research.

I thank S. N. Coppersmith, D. S. Fisher, M. P. A. Fisher, M. J. P. Gingras and P. B. Littlewood for helpful discussions.

References

[1] G. Gruner, Rev. Mod. Phys. **60**, 1129 (1988) is a review of charge-density-wave systems.

[2] L. P. Gor'kov and G. Gruner, eds., *Charge Density Waves in Solids*, (North-Holland, Amsterdam, 1989).

[3] G. Blatter, M. V. Feigel'man, V. B. Geshkenbein, A. I. Larkin and V. M. Vinokur, Rev. Mod. Phys. **66**, 1125 (1994) is a review of vortices in type-II superconductors.
[4] H. Fukuyama and P. A. Lee, Phys. Rev. B **17**, 535 (1978); P. A. Lee and T. M. Rice, Phys. Rev. B **19**, 3970 (1979).
[5] D. Feinberg and J. Friedel, J. Phys. France **49**, 485 (1988).
[6] A. I. Larkin, Zh. Eksp. Teor. Fiz. **58**, 1466 (1970) [Sov. Phys. JETP **31**, 784 (1970)].
[7] Y. Imry and S.-k. Ma, Phys. Rev. Lett. **35**, 1399 (1975).
[8] T. Nattermann, Phys. Rev. Lett. **64**, 2454 (1990).
[9] T. Giamarchi and P. Le Doussal, Phys. Rev. B **52**, 1242 (1995).
[10] D. A. Huse, Physica B **197**, 540 (1994).
[11] M. J. P. Gingras and D. A. Huse, Phys. Rev. B **53**, in press (1996).
[12] D. S. Fisher, Phys. Rev. B **31**, 1396 (1985).
[13] A. A. Middleton and D. S. Fisher, Phys. Rev. Lett. **66**, 92 (1991) and references therein.
[14] S. N. Coppersmith, Phys. Rev. Lett. **65**, 1044 (1990).
[15] S. N. Coppersmith and A. J. Millis, Phys. Rev. B **44**, 7799 (1991).
[16] H. J. Jensen, A. Brass, Y. Brechet and A. J. Berlinsky, Phys. Rev. B **38**, 9235 (1988).
[17] A.-C. Shi and A. J. Berlinsky, Phys. Rev. Lett. **67**, 1926 (1991).
[18] S. Bhattacharya and M. J. Higgins, Phys. Rev. Lett. **70**, 2617 (1993).
[19] A. E. Koshelev and V. M. Vinokur, Phys. Rev. Lett. **73**, 3580 (1994).
[20] L. Balents and M. P. A. Fisher, Phys. Rev. Lett. **75**, 4270 (1995).
[21] S. H. Strogatz, C. M. Marcus, R. M. Westervelt and R. E. Mirollo, Phys. Rev. Lett. **61**, 2380 (1988).

Large-Scale Simulations of Melting in Two-Dimensional Lennard-Jones Systems: Evidence for a Metastable Hexatic Phase

K. Chen, T. Kaplan, and M. Mostoller

Solid State Division, Oak Ridge National Laboratory,
Oak Ridge, TN 37831-6032, USA

Abstract. Large scale molecular dynamics simulations have been performed for two-dimensional Lennard-Jones systems on massively parallel computers. The calculations were done in the NPT ensemble as recently reformulated by Melchionna *et al.*, avoiding problems associated with mixed phases and artificial treatment of vacancies and interstitials. In the largest systems studied (36864 and 102400 atoms), a metastable hexatic phase was found between solid and liquid, in agreement with the predictions of the theory of melting in two dimensions developed by Kosterlitz and Thouless, Halperin and Nelson, and Young. The hexatic phase was not seen in smaller samples, calling into question the conclusions of many previous simulations performed on smaller systems and shorter time scales.

1. Introduction

In three dimensions (3D), melting is understood only on a macroscopic thermodynamic level; the phase transition is discontinuous, and requires a latent heat. There may be hysteresis, in that it is possible to superheat the solid and undercool the liquid. Empirical rules for 3D melting have been proposed, such as the Lindemann criterion advanced by its namesake 85 years ago and still in use today [1,2]. The Lindemann criterion is that a solid will melt when the root mean square amplitudes of vibrations of the atoms reach some critical value, say 15% of the interatomic spacing. This rather imprecise rule, correlating melting with the disorganized thermal vibrations of all of the atoms in the material, may be contrasted to accepted theories for other phase tranitions, e. g., the BCS electron pairing mechanism for superconductivity. At present, there is no credible, fundamental atomistic theory for melting in three dimensions.

The situation is different in two dimensions (2D). A simple and elegant theory of 2D melting mediated by topological defects was developed in a series of papers in the 1970's by Kosterlitz and Thouless [3], Halperin and Nelson [4,5], and Young [6] (KTHNY). A two-dimensional system may be characterized by two kinds of order, translational and orientational, as will be described in more detail later.

Springer Proceedings in Physics, Volume 82
Computer Simulation Studies in Condensed-Matter Physics IX
Eds.: D. P. Landau, K.K. Mon, H. -B. Schüttler

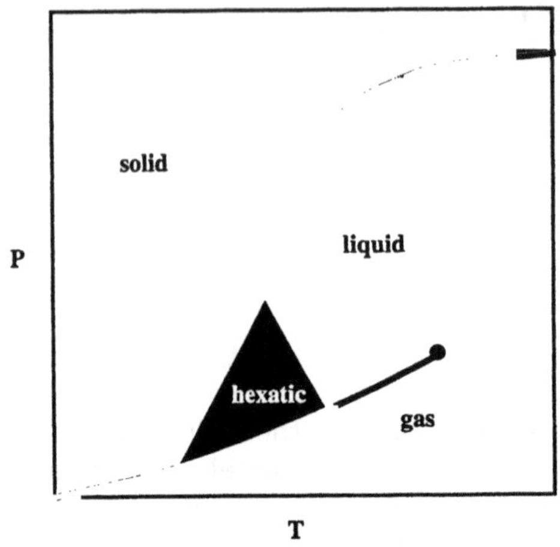

Fig. 1. Schematic NPT phase diagram for a 2D system.

A solid exhibits quasi-long-range translational order and true orientational order, while a liquid has neither. As shown in Fig. 1, the KTHNY theory proposed that in certain regions of the phase diagram, a new phase called the hexatic would exist. The hexatic phase has no long-range translational order, but retains quasi-long-range orientational order. The solid transforms into the hexatic, and the hexatic into the liquid, in two continuous phase transitions, the first by the unbinding of dislocation pairs, the second by the unbinding of pairs of disclinations. Note from the figure that the theory does not rule out the possibility of a first-order transition.

In the time since the KTHNY theory was formulated, a large number of experimental and theoretical studies have been done to try to confirm or disprove it, with ambiguous and inconclusive results. A variety of creative experiments have tended to support KTHNY, although not in every case. Murray and Van Winkle [7], for example, found evidence for a two-stage melting process through a hexatic phase in the translational and orientational correlation functions for charged polystyrene-sulphonate submicron spheres confined in water between smooth glass plates. On the other side, Pouligny *et al.*'s experiments on steel spheres confined in a parallel plate capacitor [8] suggested a first-order transition, but their system was small and uncertainties were raised by the presence of in-plane confining walls. Work by Geer *et al.* [9] on free-standing liquid crystal films and by Kusner and coworkers [10] on 2D systems of colloidal particles with dipolar interactions support the KTHNY picture.

A great many computer simulations have been performed for 2D systems with various sizes, statistical ensembles, and potential models in the past 20 or so years. Many, but not all, of these have been equivocal or negative regarding the predictions of KTHNY. For Lennard-Jones systems, Frenkel and McTague [11] did isothermal-isochoric (NVT = constant number of particles N, volume V, and temperature T) simulations for 256 atoms and claimed to find evidence for the hexatic phase. On the other hand, Abraham [12] concluded that the melting transition is first order in an early attempt to do isothermal-isobaric (constant N, pressure P, and T) simulations for 256- and 576-particle systems. NVT simulations by Toxvaerd [13] for as many as 3600 atoms, by Strandburg *et al.* [14] for systems of 1024 particles, and by Bakker *et al.* [15] for 10864 atoms led them to report first-order transitions, but Udink and van der Elsken [16], who applied finite-size scaling to results for a 12480 particle system, supported KTHNY.

For other interatomic potentials, the results are similarly mixed. Weber and coworkers [17 and references therein] have performed a thorough analysis of melting in systems of up to 16384 hard disks using finite-size scaling and the behavior of the bond-orientational susceptibilities. They provided strong evidence that melting is a first order transition in hard disk systems. At about the same time, Bladon and Frenkel [18] showed that when a short-range attractive interaction is grafted onto hard disks, a crystal becomes unstable to dislocation unbinding and the formation of a hexatic phase near a solid-solid critical point between high- and low-density solid phases.

Earlier this year, we reported [19] on molecular dynamics simulations of large 2D Lennard-Jones systems at long time scales using correct NPT equations of motion. A metastable hexatic phase was observed, but only in the two largest systems studied, providing perhaps the most direct evidence to date for the KTHNY theory of melting in two dimensions. We have since extended these calculations to do more extensive runs for the largest system.

The remainder of the paper is organized as follows: Section 2 provides a brief description of the KTHNY theory as it applies to a 2D Lennard-Jones system. Section 3 provides a few details about the simulations. Results are given in Section 4, and conclusions and future plans are outlined in Section 5.

2. Theory

The Lennard-Jones or 12-6 potential as we use it is cut off at a distance r_c and shifted so that the potential goes to zero at the cutoff,

$$V(r) = 4\varepsilon\,[(\sigma/r)^{12} - (\sigma/r)^{6}] + V_c, \qquad r < r_c, \tag{1a}$$

$$= 0, \qquad r \geq r_c, \tag{1b}$$

$$V_c = -4\varepsilon\,[(\sigma/r_c)^{12} - (\sigma/r_c)^{6}] \tag{1c}$$

There is a discontinuity in the forces and and higher derivatives of the interatomic potential at the cutoff in this model, but this appears to have little effect on the simulations in runs we have done with this potential and with closely related ones that have no discontinuities in any derivatives at the cutoff. The zero-temperature equilibrium lattice for the Lennard-Jones potential in 2D is a triangular lattice. Each atom has six first nearest neighbors (1nns) at angles spaced by 60°.

The KTHNY theory is based on an elastic continuum Hamiltonian [20]. In 2D, unlike in 3D, dislocation dipoles are thermally activated point defects; to produce finite rather than infinite elastic strain energies, dislocations must be created in pairs. Each dislocation is in turn comprised of two nearest neighbor disclinations, which are atomic sites with 5- or 7-fold coordination instead of the perfect crystalline solid coordination of 6. In the KTHNY theory, the transition from the solid to the hexatic is accomplished by the unbinding of dislocation pairs, and going from the hexatic to the liquid occurs with the unbinding of pairs of disclinations.

Spatial correlation functions are used to monitor the evolution of the system. The translational correlation function for a reciprocal lattice vectors **G** is

$$g_{\mathbf{G}}(r) = < \exp\{ i\mathbf{G}\cdot [\mathbf{u}(\mathbf{r}) - \mathbf{u}(0)]\} >, \tag{2a}$$

where **u(r)** is the atomic displacement field. For computational reasons, it is more convenient to use the pair correlation function

$$g(r) = (V/N^2) < \Sigma_i \Sigma_{j\neq i} \delta(\mathbf{r} - \mathbf{r}_{ij}) >. \tag{2b}$$

The orientational correlation function $g_6(r)$ is defined by

Table 1. Behavior of the translational and orientational correlation functions and grouping of dislocations and disclinations in the solid, hexatic and liquid phases in two dimensions.

	Solid	**Hexatic**	**Liquid**
$g_{\mathbf{G}}(r)$	$r^{-\eta(T)}$	$e^{-r/\xi(T)}$	$e^{-r/\xi(T)}$
$g_6(r)$	$g_6(\infty) \neq 0$	$r^{-\eta_6(T)}$	$e^{-r/\xi_6(T)}$
Dislocations	pairs	free	free
Disclinations	quartets	pairs	free

$$g_6(r) = < \psi_6(\mathbf{r})\, \psi_6^*(0) >, \tag{3a}$$

$$\psi_6(\mathbf{r}_i) = (1/N_{nn})\, \Sigma_{j=1nn} \exp(6i\theta_{ij}), \tag{3b}$$

where the angle θ_{ij} is defined relative to an arbitrary fixed axis. Table 1 shows how the different phases are ordered. In the solid, the long-range translational order decays algebraically, and the orientational order approaches a constant. In the hexatic, the translational correlation function decays exponentially, while the orientational order is quasi-long-range and decays algebraically. The liquid exhibits no long-range order; both correlation functions decay exponentially.

3. Computations

Most previous simulations of 2D melting have been done in an NVT ensemble. This can lead to problems associated with the existence of mixed phase regions, which make it difficult to use the correlation functions as signatures of the different phases. Furthermore, there are questions about how to maintain correct thermal equilibrium concentrations of vacancies [20, 21]. We have therefore performed all of our molecular dynamics (MD) simulations in the NPT ensemble, where such problems do not arise. The first treatment of NPT dynamics allowing variable shape as well as volume for the computational unit cell was introduced by Parrinello and Rahman (PR) [22]. It was subsequently pointed out by Cleveland and others that PR MD is not invariant to cell-basis transformations [23-25]. This was corrected in the recent reformulation of exact NPT equations of motion having the PR form by Melchionna *et al.* [26]. Using these equations allows us to make an unequivocal test of the KTHNY theory.

Simulations were performed for systems of 576, 4096, 16384, 36864, and 102400 atoms confined in parallelograms with varying cell sides as determined from the equations of motion. Periodic boundary conditions were imposed. In most cases, the starting samples were perfect triangular crystals with random velocities assigned to the atoms corresponding to the desired temperature, but in some instances hexatic or liquid starting configurationss were used. The equations of motion were integrated by the 5-value Nordsieck-Gear predictor-corrector method. In reduced units in which the potential parameters ε and σ, the atomic mass m, and the Boltzmann constant k_B are 1, the time step ranged from 0.0005 to 0.001, which gives a linear drift in the constant of motion of only 2 parts in 10^5 in the course of one million time steps. For the calculations reported here, the cutoff for the potential was $r_c = 4\sigma$.

The simulations were done on the massively parallel ORNL Intel Paragon computers XPS/35 (512 processors) and XPS/150 (2048 processors); the numerical designations correspond to nominal peak Gflops. An atom decomposition scheme was used in the parallel code. Most runs were done with 128 processors. To give a flavor for the CPU time involved, 5 x 10^6 time steps for 36864 atoms on 128 processors on the XPS/35 takes approximately 10.5 days of continuous running time, which in the real world translates to about a month. This makes it difficult to pursue every potentially good idea.

4. Results

To search for the existence of the hexatic phase, systematic scans of the P, T phase space were carried out for different system sizes. To monitor the state of the systems, the translational and orientational correlation functions were determined [see Eqs. (2) and (3)] for snapshot

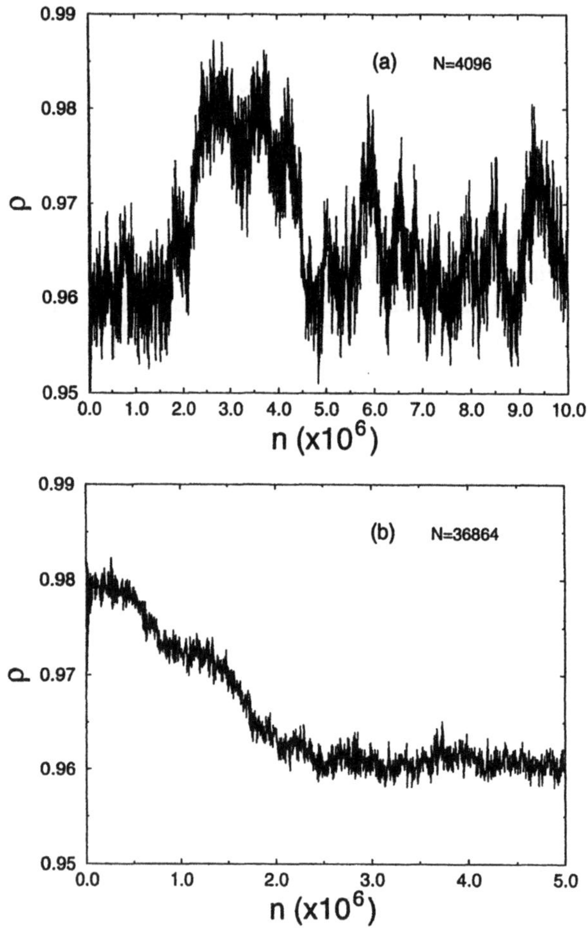

Fig. 2. Density ρ vs. time step n at P=20 and T=2.16 for 4096 atoms (a) and 36864 atoms (b). The time step Δt = 0.0005 for these runs.

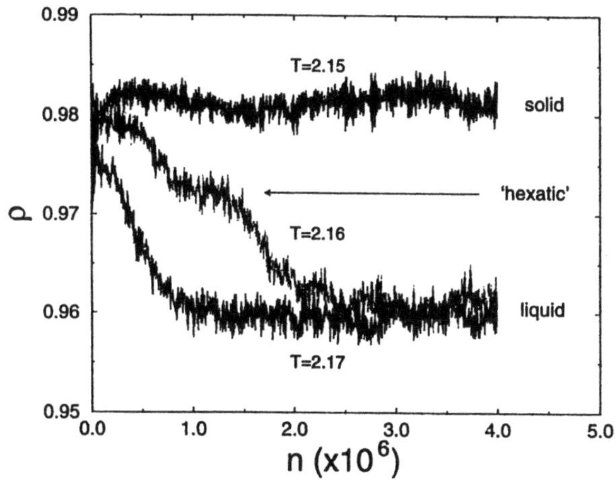

Fig. 3. Density ρ vs. time step n in a 36864-atom sample at three different temperatures near the melting point.

configurations. One feature that quickly became manifest was the effect of system size. This is illustrated in Fig. 2, which presents results for the evolution of the density ρ as a function of time for 4096- and 36864-atom periodically repeated unit cells. The 4096-atom sample was run for 10 million time steps of $\Delta t = 0.0005$, the 36864-atom unit cell for 5 million. The ordinate range is the same for the two samples. Note first that the fluctuations are much larger for the smaller sample, which apparently jumps between the liquid and solid phases in a way consistent with a first-order transition. The larger sample exhibits not only much smaller fluctuations in the density, but also a plateau in the evolution of the density from the solid at the beginning to the liquid at the end. This plateau persists for roughly 500000 time steps, and the correlation functions for this metastable state have the signature of the hexatic phase. It is apparent that large system sizes are important for meaningful simulations of 2D melting, and that results of many previous simulations on small samples are suspect.

Fig. 3 shows the behavior of the density as a function of time for the 36864-atom system at P=20 and three values of T around the melting temperature. All runs were started from the same perfect crystal lattice structure. At T=2.15, the system remains in the solid phase, while at T=2.17, it relaxes smoothly to the liquid. At the intermediate temperature T=2.16, the system relaxes first into a metastable hexatic state which persists for some time, and then on down to the liquid.

To test the sensitivity of our results to the starting configuration (and to try to stabilize the hexatic phase), runs at P=20 and T=2.154 were initiated with the perfect crystal structure and with a hexatic configuration obtained at T=2.16. The results are shown in Fig. 4. Both

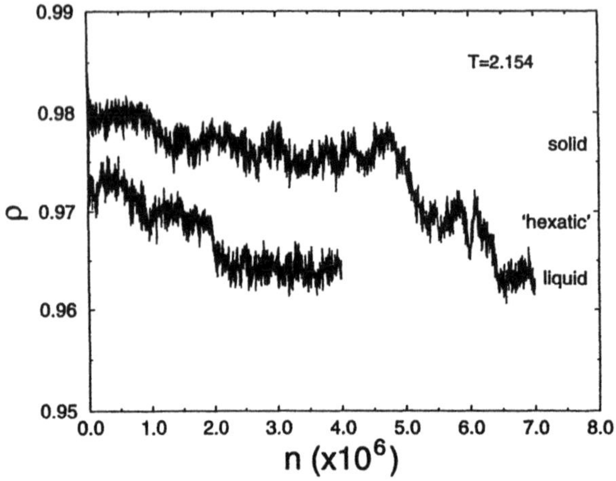

Fig. 4. Density ρ vs. time step n in a 36864-atom sample starting from the solid (top curve) and from a hexatic configuration (bottom curve).

runs confirm the existence of a metastable hexatic state, particularly the (shorter) run started with a hexatic configuration, where the metastable hexatic phase persisted for about 1 million time steps. However, neither run succeeded in stabilizing the hexatic phase.

Pair distribution and orientational correlation functions for snapshots taken from the simulations are shown in Fig. 5. At T=2.15, the system is a solid; the oscillations in the pair distribution function persist over the entire range of distance, and the orientational correlation function decays to a finite constant value. At T=2.17, the system is clearly in the liquid state, as the pair distribution decays rapidly to an average value and $g_6(r)$ decays exponentially. The system at T=2.154 displays the signature of the hexatic phase: there is only short-range order in the pair distribution function, but the orientational correlation function decays algebraically. The exponent for the decay is 0.21±0.02, which is less than the upper limit of 1/4 predicted by KTHNY. Although a mixed solid-liquid coexistence region is unlikely to occur in an NPT simulation, we have done finite-size scaling analysis of the bond-angular susceptibility in subgrids of the 36864 atom sample. All three phases observed as the temperature changes appear to be homogeneous.

System size is clearly a very important aspect of simulations of (and experiments on) 2D melting. This makes sense in the framework of KTHNY, which is based on an elastic continuum Hamiltonian for interactions between dislocations and disclinations. In the largest system discussed to this point (36864 atoms), nearly 10% of the atoms interact with periodic images of other atoms in the unit cell. We have now done a number of runs for a 320x320 sample of 102400 atoms, and while we have not succeeded in stabilizing the hexatic phase, we have extended its

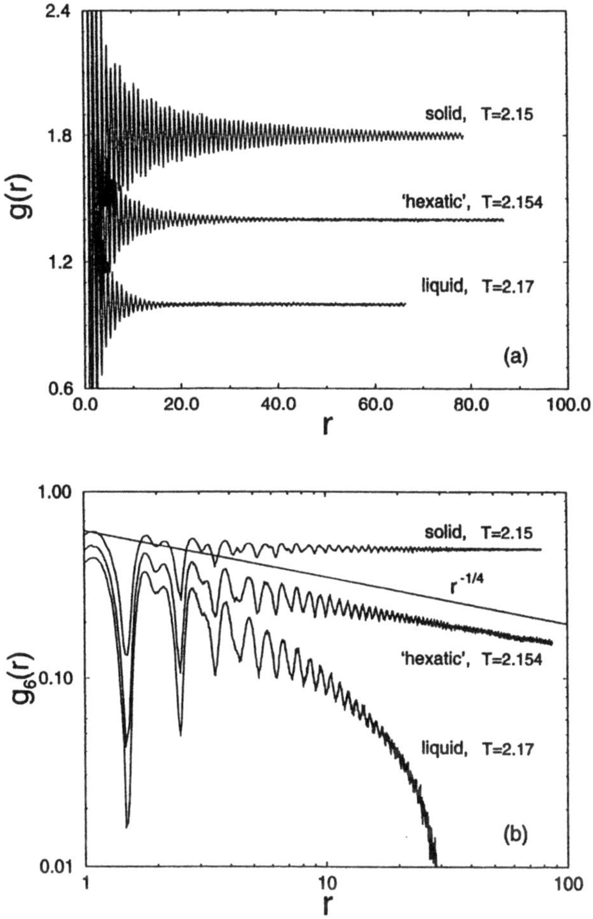

Fig. 5. Pair distribution (a) and orientational corrrelation (b) functions in a 36864-atom sample at P=20 and three temperatures near the melting point. The pair distribution functions for the solid and the metastable hexatic phases have been shifted upward for clarity.

duration as shown in Fig. 6. Simulations for the 102400 systems were started with a hexatic-like configuration, and are compared to results for the 36864-atom sample started from solid and hexatic configurations. The dashed lines are included as guides to indicate the duration of the metastable hexatic state, which persists for three million time steps in the larger system.

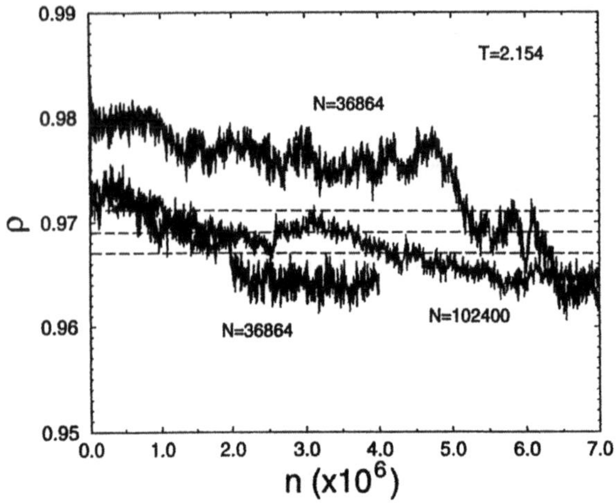

Fig. 6. Evolution of the density with time at P=20 and T=2.154 for 36ε
and 102400-atom systems.

5. Conclusions and Future Directions

We have performed extensive simulations of 2D Lennard-Jones s in a recently reformulated version of NPT dynamics. We believe t simulations provide the strongest evidence to date for the existen(hexatic phase predicted by KTHNY, and for their defect-mediate of melting in two dimensions. System size and simulation time s clearly crucial aspects of the study of the fundamental phenon melting, and present and future computer resources open vistas make it possible to provide definitive answers.

At present, we are studying the range of the Lennard-Jones its effect on dislocation core energies and the melting t Subsequent to our work [19], studies of 2D systems with purel r^{12} interactions have been published that also appear to s KTHNY theory [27]. A variety of other possibilities have bee to us, such as the imposition of external fields to stabilize the previously noted, CPU time requirements make it problemε every potentially fruitful path.

6. Acknowledgments

We are grateful to K. L. Kliewer and B. A. Riley for their g facilitating access to the Intel Paragons at the ORI Computational Sciences. Support for this work was p

ORNL Laboratory Directed Research and Development (LDRD) Program. The Oak Ridge National Laboratory is operated for the U. S. Department of Energy by Lockheed Martin Energy Systems under Contract DE-AC05-84OR21400.

References

1. A. Lindemann, Z. Phys. **11**, 609 (1910).
2. N. Q. Lam and P. R. Okamoto, MRS Bulletin/July 1994, p. 41.
3. J. M. Kosterlitz and D. J. Thouless, J. Phys. C **6**, 1181 (1973).
4. B. I. Halperin and D. R. Nelson, Phys. Rev. Lett. **41**, 121 (1978).
5. D. R. Nelson and B. I. Halperin, Phys. Rev. B **19**, 2457 (1979).
6. A. P. Young, Phys. Rev. B **19**, 1855 (1979).
7. C. A. Murray and D. H. Van Winkle, Phys. Rev. Lett. **58**, 1200 (1987).
8. B. Pouligny, R. Malzbender, P. Ryan, and N. A. Clark, Phys. Rev. B **42**, 988 (1990).
9. R. Geer, T. Stoebe, C. C. Huang, R. Pindak, G. Srajer, J. W. Goodby, M. Cheng, J. T. Ho, and S. W. Hui, Phys. Rev. Lett. **66**, 1322 (1991).
10. R. E. Kusner, J. A. Mann, J. Kerins, and A. J. Dahm, Phys. Rev. Lett. **73**, 3113 (1994).
11. D. Frenkel and J. P. McTague, Phys. Rev. Lett. **42**, 1632 (1979).
12. F. F. Abraham, Phys. Rep. **80**, 339 (1981); S. W. Koch and F. F. Abraham, Phys. Rev. B **27**, 2964 (1983).
13. S. Toxvaerd, Phys. Rev. A **24**, 2735 (1981).
14. K. J. Strandburg, J. A. Zollweg, and G. V. Chester, Phys. Rev. B **30**, 2755 (1984).
15. A. F. Bakker, C. Bruin, and H. J. Hilhorst, Phys. Rev. Lett. **52**, 449 (1984).
16. C. Udink and J. van der Elsken, Phys. Rev. B **35**, 279 (1987).
17. H. Weber, D. Marx, and K. Binder, Phys Rev. B **51**,14636 (1995); this paper contains an extensive list of references to previous work.
18. P. Bladon and D. Frenkel, Phys. Rev. Lett. **74**, 2519 (1995).
19. K. Chen, T. Kaplan, and M. Mostoller, Phys. Rev. Lett. **74**, 4019 (1995).
20. An excellent review of the general theory and earlier calculations and experiments can be found in K. J. Strandburg, Rev. Mod. Phys. **60**, 161 (1988). The total strain energy associated with an isolated dislocation diverges as log(R), where R measures the system size, but for a pair of dislocations with equal and opposite Burgers vectors, the strain energy is finite.
21. W. C. Swope and H. C. Anderson, Phys. Rev. A **46**, 4539 (1992).
22. M. Parrinello and A. Rahman, J. Appl. Phys. **52**, 7182 (1981); J. Chem. Phys. **76**, 2662 (1982).

23. C. L. Cleveland, J. Chem. Phys. **89**, 4987 (1988).
24. R. M. Wentzcovich, Phys. Rev. B **44**, 2358 (1991).
25. M. Li and W. M. Johnson, Phys. Rev. B **46**, 5237 (1992).
26. S. Melchionna, G. Cicotti, and B. L. Holian, Mol. Phys. **78**, 533 (1993).
27. K. Bagshi, H. C. Andersen, and W. Swope, Phys. Rev. Lett. **76**, 255 (1996).

Part II

Quantum Systems

Cluster Monte Carlo Method for Quantum Systems

N. Kawashima

Department of Physics, Toho University, Miyama, Funabashi,
Chiba 274, Japan

Abstract. The general framework of the cluster Monte Carlo algorithms for quantum systems is reviewed. Some new algorithms are also presented. As an example, a new algorithm is applied to the one-dimensional t-J model and has proved to reduce the autocorrelation time by a few orders of magnitude.

1 Introduction

In most practical Monte Carlo simulations, we encounter difficulties due to long auto-correlation times. We are often interested in systems near the criticality or at low temperature. These are the very situations where the slowing-down is severe. In addition, there is another source of the slowing down when one deals with quantum problems using the world-line Monte Carlo method, one of the common tools for computation of quantum systems. In the world-line quantum Monte Carlo simulation, one usually descretizes the imaginary time path-integral[1]. In order to reduce the systematic error due to this descretization, one needs to take an imaginary time spacing as small as possible. The autocorrelation time, however, diverges as the spacing decreases. In this way, a Monte Carlo simulation is almost always accompanied by the difficulty of the slowing down.

Number of attempts have been made to reduce this difficulty. For the classical system, such as the ferromagnetic Ising model, the Swendsen-Wang (SW) algorithm[2] was found to reduce the autocorrelation time near the criticality by several orders of magnitude. In the SW algorithm, one first form clusters by connecting neighboring spins stochastically, then flip each cluster with a probability 1/2. The reason for the SW algorithm's success lies in the correspondence between the clusters thus formed and "real" clusters which cause the slow relaxation. On the other hand, efforts for constructing similar algorithms for a wider range of problems had not been fully rewarded. Several examples of these efforts can be found in [4]. They are mostly on the same line as Swendsen and Wang's idea. Among the non-trivial generalization of the idea, particularly remarkable were the loop algorithm for the 6-vertex model[5] and the algorithm for the fully frustrated ising model[6]. Because of the mathematical mapping of the spin 1/2 XXZ spin models to 6-vertex models, the loop algorithm can be taken as an algorithm for the special quantum spin models. As such, an application to the spin 1/2 antiferromagnetic Heisenberg model[7] was performed successfully. We remark that the loop algorithm can be viewed as the simplest example of the general method that we present in this paper.

Springer Proceedings in Physics, Volume 82
Computer Simulation Studies in Condensed-Matter Physics IX
Eds.: D. P. Landau, K.K. Mon, H. -B. Schüttler

In this paper, we review cluster algorithms for quantum systems such as XYZ quantum spin models with arbitrary magnitude of spins [8] and $t-J$ models[9]. We place the foundation of these algorithms on the generalized Fortuin-Kasteleyn (FK) representation[3], which is the mathematical basis of the SW algorithm. In that way, we naturally arrive at the cluster algorithms of various systems under consideration.

2 Generalized Fortuin-Kastelyn representation

We first present the general framework which is common to all existing cluster algorithms which we know of. The discussion in this section is not specific to quantum systems, but apply equally well to any stochastic system which is not even necessarily a statistical mechanical problem. The general framework of cluster algorithms was first discussed by Kandel and Domany[6]. We represent it here in an even more general form [10] which is suitable for the present purpose.

Most of interesting many-body problems in statistical machanics are defined in terms of few-body interactions among elements. It means that we can factorize the weight function $W(S)$ into a product of factors $w(S(u))$, i.e.,

$$W(S) \equiv \prod_{u} w(S(u)), \tag{1}$$

where $S(u)$ is the restriction of the state S to the set of a few elements u and $w(S(u))$ represents the few-body interaction among elements in u. We call u ***a local unit*** because u represents a set of elements near to each other in a typical application where distance is defined. In this paper, we assume the functional form of w does not depend on the local unit u for the simplicity's sake, though the generalization to non-uniform cases is trivial.

Here we introduce an auxiliary variable $G(u)$ on each local unit, and also define a delta function $\Delta(S(u), G(u))$ which takes two values 0 and 1. We call $G(u)$ ***a local graph*** because it indeed corresponds to a graph in a typical application. Although we can define these things as we like, the efficiency of the resulting algorithm strongly depends on those. We will present some guidance for choosing appropreate $\Delta(S(u), G(u))$ in what follows. The weight of the graph $v(G(u))$ is defined as a solution to the equation

$$w(S(u)) = \sum_{G(u)} \Delta(S(u), G(u)) v(G(u)). \tag{2}$$

For the reason we will see below, we must impose a requirment on the delta function that the above equation should have a solution in which all v's are non-negative. We then define a global graph, the global delta function, and the global graph weight as

$$G \equiv \bigcup_{u} G(u), \ \ \Delta(S, G) \equiv \prod_{u} \Delta(S(u), G(u)), \ \ V(G) \equiv \prod_{u} v(G(u)) \tag{3}$$

With these definitions, we can reexpress the weight function becomes

$$W(S) = \sum_G W(S,G) \tag{4}$$

where

$$W(S,G) \equiv \Delta(S,G)V(G) \tag{5}$$

and consequently the generating function as

$$Z = \sum_{S,G} W(S,G) \tag{6}$$

Since $W(S,G)$ is factorized into factors depending only on G and a delta fanction taking only 0 and 1, it is often possible to evaluate the sum over S in (6) with a given graph G. Namely,

$$Z = \sum_G V(G)N(G) \tag{7}$$

where $N(G)$ is the number of states which are compatible with G, i.e., the number of S's for which $\Delta(S,G) = 1$. This expression in the case of the Ising model is called the FK representation. For the purpose of numerical simulation, however, we will use the intermediate expression (6) rather than the FK representation (7).

We can interpret $W(S,G)$ in (5) as the weight function of the product space that consists of the original state space and the new graph space. The equation (4) implies that if we generate a pair of S and G stochastically with the weight $W(S,G)$, the distribution of the state S thus generated is $W(S)$. Therefore, instead of constructing a Markov process in the state space, we might as well construct a process in the product space. Then, we can regain the information concerning the state space only by neglecting the graphs. We then consider a Markov process on the product space such that each step consists of two sub-steps. In the first sub-step, only the graph G in the pair (S,G) is updated while in the second the state S is updated. Of course, we require the limiting distribution of the process to be proportional to $W(S,G)$. The most natural choice of transition probabilities satisfying such a condition is the one which is simply proportional to the weight of the final state:

$$P_L((S',G')|(S,G)) = \delta(S',S)W(S,G') \Big/ \sum_{G''} W(S,G'') \tag{8}$$

$$P_F((S',G')|(S,G)) = \delta(G',G)W(S',G) \Big/ \sum_{S''} W(S'',G) \tag{9}$$

where the labeling probability P_L characterizes the first sub-step (the labeling process), and the flipping probability P_F the second (the flipping process). It is easy to see that the weight function $W(S,G)$ is stable in this Markov process, which together with the ergodicity condition guarantees that the limiting distribution is proportional to $W(S,G)$. It is also not difficult to verify that the SW algorithm is of this type of Markov process[8].

To see what we should do in practice, let us examine the above two transition probabilities in more detail. Clearly, the labeling probability factorizes into local factors.

$$P_L((S',G')|(S,G)) = \delta(S',S)\prod_u P_L(G'(u)|S(u)) \tag{10}$$

where

$$P_L(G(u)|S(u)) = \frac{\Delta(S(u),G(u))v(G(u))}{w(S(u))}. \tag{11}$$

This means that the labeling process of a Markov step can be done simply by assigning a local graph $G(u)$ with the probability $P_L(G(u)|S(u))$ to each local unit u and then taking the union $\bigcup_u G(u)$ as G. As for the flipping probability, we can re-express it as

$$P_F((S',G')|(S,G)) = \delta(G',G)\frac{\Delta(S',G)}{N(G)}. \tag{12}$$

This means that we should choose one of the states compatible with G at random with equal probability. In most applications, after the labeling process, the system can be devided into clusters and we can define a cluster variable on each cluster such that a state compatible with a given graph corresponds to a cluster-variable configuration. In such cases, the flipping process reduces to choosing the cluster variables at random.

3 Quantum Spin Systems

In this section, we describe cluster algorithms[8] for quantum spin systems. In what follows, we do not assume any particular dimensionality, lattice structure, magnitude of spins, or anisotropy of the model.

The XYZ Hamiltonian we consider is

$$\mathcal{H} = -\sum_{(i,j)}(J_0 + J_x S_i^x S_j^x + J_y S_i^y S_j^y + J_z S_i^z S_j^z) \quad (\boldsymbol{S}_i^2 = S(S+1)). \tag{13}$$

Although the constant J_0 is physically irrelevant, we include it to make the subsequent discussion look simpler. First, we "divide" each spin into $2S$ Pauli operators:

$$S_i^\alpha = \frac{1}{2}\sum_{\mu=1}^{2S}\sigma_{i,\mu}^\alpha \quad (\alpha = x,y,z). \tag{14}$$

Note that we are expanding the Hilbert space by replacing the original $(2S+1)$ dimensional spin space for each spin by the 2^{2S} dimensional space for $2S$ Pauli spins. As the representation basis for the extended Hilbert space, we take simultaneous eigenfunctions of the z-components of all the Pauli matrices.

We map the original d dimensional problem into $(d+1)$ dimensions using the Suzuki-Trotter (ST) decomposition[1]. The resulting problem is a system with

$2S$ classical one-bit degrees of freedom defined on each lattice point. Accordingly, we consider $2S$ vertices on each lattice point as tools for visualization. We refer to the set of all the one-bit variables as S. The generating function of the problem can be written as

$$Z = \sum_{S} \prod_{p} \mathrm{sgn}(S(p)) w(S(p)) \tag{15}$$

where $S(p)$ is the restriction of S to the plaquette p. Here, a plaquette p is identified with the set of four lattice sites, or $8S$ vertices on which an $8S$-body interaction is defined. A plaquette, of course, corresponds to the local unit u in the general framework in the last section.

Although the sign $\mathrm{sgn}(S(p))$ in (15) is very important from the physical point of view, it is irrelevant from the computational point of view as long as one uses the conventional method of neglecting the sign in constructing the Markov process. The function $w(S(p))$ is the absolute value of the matrix element of an operator $\hat{\rho}$ defined by

$$\hat{\rho} \equiv \hat{P} e^{\hat{\Lambda}} \hat{P}, \quad \hat{\Lambda} \equiv \sum_{\mu,\nu} \hat{\Lambda}_{\mu,\nu}, \tag{16}$$

where $\hat{P}$ is the projection operator to the space where $\boldsymbol{S}_i^2 = \boldsymbol{S}_j^2 = S(S+1)$, and

$$\hat{\Lambda}_{\mu,\nu} \equiv K_0 + K_x \sigma_{i,\mu}^x \sigma_{j,\nu}^x + K_y \sigma_{i,\mu}^y \sigma_{j,\nu}^y + K_z \sigma_{i,\mu}^z \sigma_{j,\nu}^z, \tag{17}$$

with some given constants K_α ($\alpha = x, y, z$) which are proportional to the imaginary time spacing $\Delta\tau$.

We then define local graphs to be something consisting of a few colored edges. Each edge connects two vertices on a plaqutte under consideration and has its own color, green or red. The real meaning of a graph is given by the following definition of the delta function.

$$\Delta(S(p), G(p)) \equiv \begin{cases} 1 & \text{(If every green edge connects parallel spins and} \\ & \text{every red edge connects anti-parallel spins)} \\ 0 & \text{(Otherwise)} \end{cases} \tag{18}$$

In the labeling process, we assign a graph $G(p)$ to each plaquette with probability discussed below. We then define a cluster as a set of vertices connected to each other by a series of edges. As a result, the whole system is devided into clusters which have no overlap with each other. Obviously, these clusters carry the same information as the graph $G \equiv \bigcup_p G(p)$ except for the color of edges. If we consider a one-bit cluster variable defined on each cluster, there is a one-to-one correspondence between configurations of those cluster variables and states compatible with the graph $G \equiv \bigcup_p G(p)$. Therefore, as predicted in the last section, the flipping process is carried out simply by choosing cluster variables with equal probability at random, and mapping it back into a original spin configuration.

We found[8] that in general a set of non-negative graph weights $v(G(p))$ exists that satisfies (2) with u replaced by p and with the definitions of $w(S(p))$

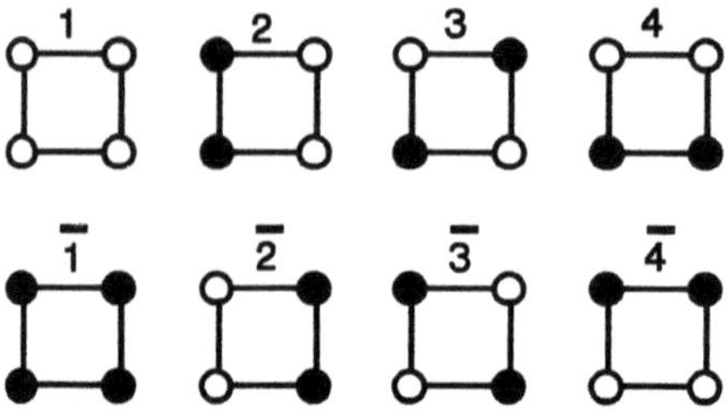

Figure 1: The eight possible local configurations on a set of four vertices, one chosen from each corner of a plaquette. Open and closed circles represent 'up' spins and 'down' spins, respectively.

and graphs mentioned above. The equation is equivalent to the following operator equation

$$|\hat{\rho}| = \sum_{G(p)\in\Gamma} v(G(p))\hat{\Delta}(G(p)). \tag{19}$$

Here, $|\hat{\rho}|$ is the operator whose matrix elements are the absolute values of those of $\hat{\rho}$. The operator $\hat{\Delta}(G(p))$ is the operator whose matrix elements are $\Delta(S(p), G(p))$.

It was shown[8] that the problem of finding a solution to (19) is reduced to the problem of finding non-negative coefficients $a(G(p))$ that satisfy

$$|\hat{\Lambda}_{\mu,\nu}| = \sum_{G(p)\in\Gamma_{\mu,\nu}} a(G(p))\hat{\Delta}(G(p)). \tag{20}$$

Here, the set $\Gamma_{\mu,\nu}$ is the set of graphs defined on only four vertices, one from each corner of the plaquette. Two of them on the left side of the plaquette correspond to the μ-th Pauli spin which is an element of $\boldsymbol{S}_i$ and the other two on the right side correspond to the ν-th Pauli spin which is an element of $\boldsymbol{S}_j$.

In Fig. 1, eight possible spin configuration which correspond to eight kinds of non-zero matrix elements of $|\hat{\Lambda}_{\mu,\nu}|$ are presented. The ten graphs in $\Gamma_{\mu,\nu}$ are shown in Fig. 2.

In the case where $K_x = K_y$, the solution to (20) is given by

$$|\hat{\Lambda}_{\mu,\nu}| = \begin{cases} 2(K_z - K_x)\hat{\Delta}_{11} + (K_0 - K_z)\hat{\Delta}_{12} + 2K_x\hat{\Delta}_{13} & (0 < K_x < K_z) \\ (K_0 - K_x)\hat{\Delta}_{12} + (K_x + K_z)\hat{\Delta}_{13} + (K_x - K_z)\hat{\Delta}_{14} & (0 < |K_z| < K_x) \\ 2(-K_z - K_x)\hat{\Delta}_{22} + (K_0 + K_z)\hat{\Delta}_{12} + 2K_x\hat{\Delta}_{14} & (0 < K_x < -K_z) \end{cases} \tag{21}$$

In the general case where K_x is not necessarily equal to K_y, it is convenient to express the operator $|\hat{\Lambda}_{\mu,\nu}|$ as 4×4 matrices of the form

$$\begin{pmatrix} K_1 & 0 & 0 & K_4 \\ 0 & K_2 & K_3 & 0 \\ 0 & K_3 & K_2 & 0 \\ K_4 & 0 & 0 & K_1 \end{pmatrix} \qquad (K_\sigma \geq 0). \tag{22}$$

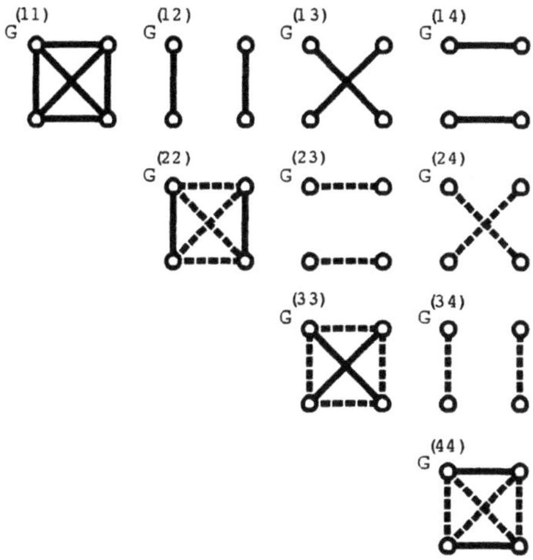

Figure 2: The ten graphs on the same set of four vertices. 'Green' edges are represented by solid lines whereas 'red' ones by dashed lines.

It can be seen that there are eight non-zero elements corresponding to eight possible states in Fig.1. Then, the equation (20) reduces to

$$K_\sigma = \sum_{\rho=1}^{4} a_{(\sigma\rho)} \qquad \left(a_{(\sigma\rho)} \equiv a(G^{(\sigma\rho)})\right), \tag{23}$$

where graphs $G^{(\sigma\rho)}$ are defined in Fig. 2. Note here that if one regards the matrix elements K_σ above as the vertex weight of an eight-vertex model, the following argument will offer a cluster algorithm for an eight-vertex model. If one regards K_σ as

$$K_1 = K_0 + K_z, \quad K_2 \equiv K_0 - K_z, \quad K_3 = |K_x + K_y|, \quad \text{and} \quad K_4 \equiv |K_x - K_y|,$$

the following can be viewed as the solution for (20). (Note that we can take K_0 as large as we like to make K_1 and K_2 positive.)

We can assume $K_1 \geq K_2 \geq K_3 \geq K_4 \geq 0$ without the loss of generality, because once we get solution for this case, we can get solutions for other cases simply by permuting indices. We then consider the following three cases: 1) $K_2+K_3+K_4 \leq K_1$, 2) $K_2+K_3-K_4 \leq K_1 \leq K_2+K_3+K_4$, and 3) $K_1 \leq K_2+K_3-K_4$. (In the second interpretation of K_σ's in the last paragraph, this means that 1) $|K_z|$ is the largest among $|K_x|, |K_y|$ and $|K_z|$, 2) $|K_z|$ is the second largest among three, and 3) $|K_z|$ is the smallest among three, respectively.)

In general, we should minimize the weights for $G^{(\sigma\sigma)}$ which are proportional to the 'freezing probability'. However, in the case 1), it is easy to see that no solution exists for which $a_{(11)}$ is zero. This means that a loop algorithm solution does not exist. The best we can seek, therefore, is a solution in which

$a_{(22)} = a_{(33)} = a_{(44)} = 0$. Among such solutions, the one that minimizes $a_{(11)}$ is

$$a_{(11)} = K_1 - K_2 - K_3 - K_4, \; a_{(1\sigma)} = K_\sigma \text{ (for } \sigma = 2,3,4), \tag{24}$$

with all other a's vanishing.

In cases 2) and 3), as we will see below, solutions exist for which all $a_{(\sigma\sigma)}$'s are zero. This kind of solution corresponds to a loop algorithm in which the clusters formed do not have any branching. In what follows, we will consider such solutions only, although other solutions also exist. There are 6 independent variables subject to the 4 independent equations (23). Therefore, the solution space is in general two-dimensional. In the matrix form, the general solution of (23) for the case 2) and 3) must have the following form:

$$A = A_0 + pP + qQ,$$

$$P \equiv \begin{pmatrix} 0 & -1 & 0 & 1 \\ -1 & 0 & 1 & 0 \\ 0 & 1 & 0 & -1 \\ 1 & 0 & -1 & 0 \end{pmatrix}, \quad Q \equiv \begin{pmatrix} 0 & -1 & 1 & 0 \\ -1 & 0 & 0 & 1 \\ 1 & 0 & 0 & -1 \\ 0 & 1 & -1 & 0 \end{pmatrix},$$

and $A_0 \equiv \frac{1}{2} \times$

$$\begin{pmatrix} 0, & 2K_2, & K_1 - K_2 + K_3 - K_4, & K_1 - K_2 - K_3 + K_4 \\ & 0, & 0, & 0 \\ & & 0, & -K_1 + K_2 + K_3 + K_4 \\ & & & 0 \end{pmatrix}$$

where the matrix A is defined by $(A)_{\sigma\rho} \equiv a_{(\sigma\rho)}$. Since A_0 is the symmetric matrix, only the upper half of it has been shown. The condition $a_{(\sigma\rho)} \geq 0$ imposes a restriction on the range of (p, q). In the case 2), we have

$$0 \leq p, \quad 0 \leq q, \quad p + q \leq \frac{1}{2}(-K_1 + K_2 + K_3 + K_4). \tag{25}$$

Similarly, in the case 3), defining $A_1 \equiv A_0 + (1/2)(-K_1 + K_2 + K_3 - K_4)A$ and $p' \equiv p - (1/2)(-K_1 + K_2 + K_3 + K_4)$, the general solution can be written as

$$A = A_1 + p'A + qB \qquad (0 \leq p', \quad 0 \leq q, \quad p' + q \leq K_4). \tag{26}$$

Thus, we have obtained solutions that covers the entire range of parameters K_σ.

4 $t - J$ Model

The hamiltonian for the $t - J$ model is given by

$$\mathcal{H} = \sum_{(ij)} \mathcal{H}_{ij} \tag{27}$$

where $\mathcal{H}_{ij}$ is an operator acting on a 3×3 local Hilbert space. A basis vector of this 9-dimensional Hilbert space is represented by a pair of two variables S_i and S_j. Each variable takes three values, 1, 0 and -1. Symbolically, the pair Hamiltonian is usually written as

$$\mathcal{H}_{ij} = -t \sum_{\sigma} (c_{i\sigma}^{\dagger} c_{j\sigma} + c_{i\sigma} c_{j\sigma}^{\dagger}) + J \boldsymbol{S}_i \boldsymbol{S}_j \tag{28}$$

The local weight $w(S(p))$ is defined as the absolute value of the matrix element of the operator $e^{-\Delta\tau\mathcal{H}_{ij}}$ with the imaginary time spacing $\Delta\tau$.

Now we consider the cluster algorithm of this model[9]. We again map the original d dimensional system into a $(d+1)$ dimensional classical system as we did in the last section. We devide the one Monte Carlo step into three steps such that in the first, second and third steps, variables with the values $-1, 0$ and 1, respectively, are left unaffected. For example, in the second step, we leave the variables with the value 0 unaffected and attempt to flip all the other variables. Here, "flip" means switch the value -1 into $+1$ and vice versa. Therefore, in each step, an active variable is effectively a binary (Ising) variable. To this binary system, we apply the idea of the cluster algorithm, which means we further devide each one of the three steps into two sub steps. The first step is the labeling process and the second is the flipping process as we discussed in Section 2. Since we have three different binary systems, we need to construct three cluster algorithms, although the first one is transformed into the third one simply by interchanging the roles of the values $+1$ and -1 for each variable S_i.

We first consider the second algorithm, i.e., the one that does not change vertices with the value 0. The algorithm in this case is essentially equivalent to

S(p)	G(p) v / w	$e^{-\Delta J/8}$	$e^{\frac{\Delta J}{8}} \mathrm{sh}\frac{\Delta J}{4}$	$\mathrm{ch}\Delta t$	$\mathrm{sh}\Delta t$	1
↑ ↑ / ↑ ↑	$e^{-\Delta J/8}$	1				
↑ ↓ / ↑ ↓	$e^{\frac{\Delta J}{8}} \mathrm{ch}\frac{\Delta J}{4}$	1	1			
↓ ↑ / ↑ ↓	$e^{\frac{\Delta J}{8}} \mathrm{sh}\frac{\Delta J}{4}$		1			
↑ o / ↑ o	$\mathrm{ch}\Delta t$			1		
o ↑ / ↑ o	$\mathrm{sh}\Delta t$				1	
o o / o o	1					1

Figure 3: The delta function $\Delta(S(p), G(p))$ for the second step in which the vertices with the value 0 are left unaffected. The graphs are shown on the top row with the graph weights $v(G(p))$ beneath them. The solid circles represent fixed vertices, i.e., the ones with the value 0 (o) in the present case.

the one for the antiferromagnetic Heisenberg model in the last section. In this case, the states of a plaquette, i.e., the set of four variables on the four corners of the plaquette, are classified into 6 classes according to the weight $w(S(p))$. In the first colomn of Fig. 3, a local state in each class is shown together with the graph weight. Local graphs are classified into 5 classes as shown in the top row. Two graphs which can be transformed to each other by an operation such as a reflection with respect to a vertical or horizontal axis belong to the same class. It is easy to veryfy that the graph weight $v(G(p))$ shown in Fig. 3 satisfies the weight equation (2). Then, it is straightforward to compute the labeling probability using the formula (11).

As for the first algorithm, i.e., the one that does not change vertices with the value -1, we have several alternatives as we see below. We will present two such alternatives. In both of them, we have to devide the local weight into two factors.

$$w(S(p)) = w_L(S(p))w_F(S(p)) \tag{29}$$

The first one will be taken into account in the labeling probability while the second in the flipping probability. Here we are modifying the general scheme presented in Section 2 where the flipping probability of each cluster was always 1/2. In the present algorithm, $w(S(p))$ in (2) is replaced by $w_L(S(p))$. In order to take care of the remnant of the weight, i.e., $w_F(S(p))$, we change the flipping probability of a cluster from 1/2 to $P_{\text{flip}} = W_F(S')/(W_F(S)+W_F(S'))$ where $W_F(S) \equiv \prod_p w_F(S(p))$, S is the initial state, and S' is the final state after flipping the cluster. Note that clusters are in general interacting, i.e., the flipping of a cluster changes the probability of flipping another one. Although this is the drawback in this approach, the clusters are independent in the second alternative as we will see below.

We do not have a strong guiding principle for how the division in (29) should be made. We will make it so that the resulting algorithm is as simple as possible. In the first alternative of the algorithm, we set

$$w_L(S(p)) = \begin{pmatrix} e^{-\Delta t} & & & & & & & & \\ & 1 & & & & & & & \\ & & 1 & 1 & & & & & \\ & & 1 & 1 & & & & & \\ & & & & \mathrm{ch}\Delta t & \mathrm{sh}\Delta t & & & \\ & & & & \mathrm{sh}\Delta t & \mathrm{ch}\Delta t & & & \\ & & & & & & 1 & 1 & \\ & & & & & & 1 & 1 & \\ & & & & & & & & e^{-\Delta t} \end{pmatrix} \begin{matrix} \uparrow\uparrow \\ \downarrow\downarrow \\ \uparrow\downarrow \\ \downarrow\uparrow \\ \uparrow\circ \\ \circ\uparrow \\ \downarrow\circ \\ \circ\downarrow \\ \circ\circ \end{matrix} \tag{30}$$

where $\Delta t \equiv t\Delta\tau$. Here, the last column represents the basis vector of the representation. For example, $\uparrow\downarrow$ represents the state $(S_i, S_j) = (+1, -1)$ and $\uparrow\circ$ represents $(+1, 0)$. This representation simplifies the labeling process since the weight $w_L(S(p))$ is similar to that for the $S = 1/2$ antiferromagnetic Heisenberg model if we focus on the matrix elements concerning the active values, namely, $+1(\uparrow)$ and $0(\circ)$. We can therefore construct an algorithm similar to that for

S(p) \ w_L \ v		$e^{-\Delta t}$	$\mathrm{sh}\,\Delta t$	1	1	1
↑ ↑ / ↑ ↑	$e^{-\Delta t}$	1				
o o / o o	$e^{-\Delta t}$	1				
↑ o / ↑ o	$\mathrm{ch}\,\Delta t$	1	1			
o ↑ / ↑ o	$\mathrm{sh}\,\Delta t$		1			
↑ ↓ / ↑ ↓	1			1		
o ↓ / o ↓	1			1		
↓ ↑ / ↑ ↓	1				1	
↓ o / o ↓	1				1	
↓ ↓ / ↓ ↓	1					1

Figure 4: The delta function $\Delta(S(p), G(p))$ for the first alternative algorithm of the first step in which the vertices with $\downarrow$ are left unaffected. The weights $w_L(S(p))$ taken into account in the labeling process are shown on the right of the corresponding state (the leftmost column).

the Heisenberg model which is characterized by the delta function. The delta function characterizing the algorithm is shown in Fig. 4.

For the second alternative of the algorithm, we set

$$w_L(S(p)) = \begin{pmatrix} e^{-\frac{\Delta J}{8}} & & & & & & & & \\ & 1 & & & & & & & \\ & & 1 & 1 & & & & & \\ & & 1 & 1 & & & & & \\ & & & & \mathrm{ch}\Delta t & \mathrm{sh}\Delta t & & & \\ & & & & \mathrm{sh}\Delta t & \mathrm{ch}\Delta t & & & \\ & & & & & & 1 & 1 & \\ & & & & & & 1 & 1 & \\ & & & & & & & & e^{-\frac{\Delta J}{8}} \end{pmatrix} \begin{matrix} \uparrow\uparrow \\ \downarrow\downarrow \\ \uparrow\downarrow \\ \downarrow\uparrow \\ \uparrow\circ \\ \circ\uparrow \\ \downarrow\circ \\ \circ\downarrow \\ \circ\circ \end{matrix} \qquad (31)$$

where $\Delta J \equiv J\Delta\tau$. Accordingly, the solution is shown in Fig. 5 with Table 1. We have to use more graphs for this altenative than for the first one. At the cost of this complication, the algorithm gains the property that we can flip each cluster independently although the flipping probability is not necessarily 1/2. In order that the flipping can be done independently, we must require that for any graph $G(p)$ which consists of two seperate edges e_1 and e_2, $w_F(S(p))$ factorizes as follows

$$w_F(S(p)) = w'_F(S(e_1))w'_F(S(e_2)) \qquad (32)$$

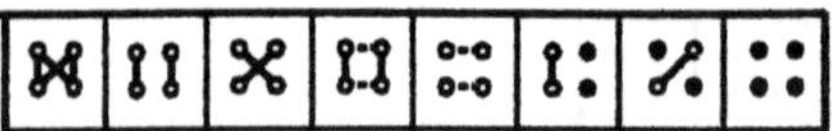

S(p) \ G(p), w_L \ v		v_{11}	v_{12}	v_{13}	v_{22}	v_{23}	1	1	1
↑ ↑ / ↑ ↑	$e^{-\Delta J/8}$	1	1	1					
∘ ∘ / ∘ ∘	$e^{-\Delta J/8}$	1	1	1					
↑ ∘ / ↑ ∘	chΔt		1		1	1			
∘ ↑ / ↑ ∘	shΔt			1		1			
↑ ↓ / ↑ ↓	1						1		
∘ ↓ / ∘ ↓	1						1		
↓ ↑ / ↑ ↓	1							1	
↓ ∘ / ∘ ↓	1							1	
↓ ↓ / ↓ ↓	1								1

Figure 5: The delta function $\Delta(S(p), G(p))$ for the second alternative algorithm of the first step.

Table 1: The graph weights which appear in Fig.5.

	$J < -8t$	$-8t < J < 8t$	$8t < J$
v_{11}	$e^{-\Delta J/8} - e^{\Delta t}$	0	0
v_{12}	$\cosh \Delta t$	$\frac{e^{-\Delta J/8}+e^{-\Delta t}}{2}$	$e^{-\Delta J/8}$
v_{13}	$\sinh \Delta t$	$\frac{e^{-\Delta J/8}-e^{-\Delta t}}{2}$	0
v_{22}	0	0	$e^{-\Delta t} - e^{-\Delta J/8}$
v_{23}	0	$\frac{-e^{-\Delta J/8}+e^{\Delta t}}{2}$	$\sinh \Delta t$

with some function w'_F, as far as the states compatible with $G(P)$ are concerned. For example, for the graph with two vertical edges (i.e., the second graph from the left in the top row of the Fig. 5), there are four compatible states, i.e.,

$$\begin{pmatrix} \uparrow & \uparrow \\ \uparrow & \uparrow \end{pmatrix}, \quad \begin{pmatrix} \uparrow & \circ \\ \uparrow & \circ \end{pmatrix}, \quad \begin{pmatrix} \circ & \uparrow \\ \circ & \uparrow \end{pmatrix}, \quad \text{and} \quad \begin{pmatrix} \circ & \circ \\ \circ & \circ \end{pmatrix}, \tag{33}$$

corresponding to the weights $e^{-\Delta J/8}$, 1, 1 and $e^{\Delta J/8}$, respectively. If we define

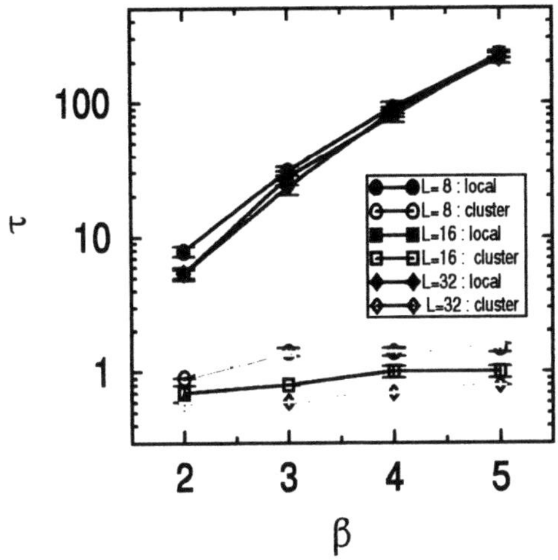

Figure 6: The integrated auto-correlation time of the one-dimensional $t-J$ model measured with the susceptibility plotted against the inverse temperature β. The ratio of J and t is $J/t = 2$. The particle density is $n = 1/2$.

e_1 to be the left edge and e_2 the right one, and if we let the variable at the lower-left vertex represent the state of the left edge and the variable at the lower-right vertex the state for the right edge, the above four states can be identified with a pair of two such variables, S_L and S_R. Then, it is easy to veryfy that the weights for the four states can be expressed as $w'_F(S_L)w'_F(S_R)$ with $w'_F(1) = \exp(-\Delta J/16)$ and $w'_F(0) = \exp(\Delta J/16)$.

To see the efficiency of the new algorithm, in Fig. 6, we show the autocorrelation time for the susceptibility at half-filling ($\langle n \rangle = 0.5$) for various system sizes as a function of the inverse temperature $\beta \equiv 1/T$. For this figure, we used the first alternative. We can clearly observe that the correlation time increases as the temperature decreases for the conventional algorithm while the correlation time for the cluster algorithm remains of the order of unity regardless of the system size and the temperature.

5 Summary

We have reviewed the general framework of the cluster algorithm. It is clarified that the cluster algorithm for the quantum system can be constructed in the same fashion as the classical system and that the quantum apllication is as powerful as the classical one. However, we still do not know exactly what makes the cluster algorithm effective. In the case of the SW algorithm for the Ising models, one can point out that the efficiency is related to the fact that the average size of the cluster is proportional to the susceptibility. However, for the

$S = 1/2$ XY model, for instance, the cluster algorithm is still efficient although there is no such relationship between the cluster size and the susceptibility. In order to construct powerful algorithms for various other yet uncovered problems such as boson systems and systems with continuous degrees of freedom, it would be helpful to study the reason for the remarkable efficiency (or inefficiency sometimes) of the existing cluster algorithms.

I would like to thank the Inoue foundation for the financial support.

References

[1] M. Suzuki, *Prog. Theor. Phys.* **56**, 1454 (1976).

[2] R. H. Swendsen and J.-S. Wang, *Phys. Rev. Lett.* **58**, 86 (1987).

[3] P. W. Kasteleyn and F. M. Fortuin, *J. Phys. Soc. Jpn.* **26** Suppl., 11 (1969); C. M. Fortuin and P. W. Kasteleyn, *Physica* **57**, 536 (1972).

[4] R. H. Swendsen, J.-S. Wang and A. M. Ferrenberg, in *The Monte Carlo Method in Condensed Matter Physics*, K. Binder ed. (Springer, Berlin, 1991).

[5] H. G. Evertz, G. Lana and M. Marcu, *Phys. Rev. Lett.* **70**, 875 (1993).

[6] D. Kandel and E. Domany, *Phys. Rev. B* **43** 8539 (1991).

[7] U.-J. Wiese and H.-P. Ying, *Phys. Lett. A* **168** 143 (1992).

[8] N. Kawashima and J. E. Gubernatis, *Phys. Rev. Lett.* **73**, 1295 (1994); N. Kawashima and J. E. Gubernatis, *J. Stat. Phys.* **80**, 169 (1995); N. Kawashima, *J. Stat. Phys.* **82** 131 (1996).

[9] B. Ammon, H. G. Evertz, N. Kawashima and M. Troyer, in preparation.

[10] N. Kawashima and J. E. Gubernatis, *Phys. Rev.* E **51** 1547 (1995).

Coupled Quantum Chains: A Bumpy Path from 1 to 2 Dimensions

H.G. Evertz

Theor. Physik, Univ. Würzburg, D-97074 Würzburg, Germany
evertz@physik.uni-wuerzburg.de

Abstract. A review of the current theoretical knowledge on coupled quantum chains is given, with emphasis on numerical results. These systems occur in the same materials as high T_c superconductors. They allow the study of the dimensional crossover from 1 to 2 dimensions. The behavior of coupled chains, even at isotropic couplings, can be understood in the rung-picture, i.e. the case of large perpendicular couplings. Undoped systems correspond to the Heisenberg spin $\frac{1}{2}$ antiferromagnet, and are well understood. They possess a finite spin gap for an *even* number of coupled chains, and are gapless otherwise. Doped systems are much more difficult to investigate, and only systems with two and three coupled chains have been partially studied.

1 Introduction

Coupled Quantum Chains, also called ladder systems, have been the topic of a lot of recent experimental and theoretical research. This paper presents an overview of the current theoretical understanding of their static behavior, concentrating mainly on numerical studies. It does not attempt to fully cover the literature on the subject. For an excellent review including experimental findings, see [1].

There are several reasons for the theoretical interest in ladder systems. First, real ladder materials do exist, mostly in compounds related to High Temperature Superconductors. Second, ladder systems provide a good way to study the dimensional crossover from 1 to 2 dimensions; this will hopefully also lead to a better understanding of high T_c superconductivity. Third, coupled chains are much easier to investigate, especially numerically, than two-dimensional systems, and one can make use of the extensive knowledge about single chains. Fourth, there is a very interesting analogue of the Haldane conjecture of single chain fame (see below).

The main messages of this paper: The intuitively easy *rung picture* ($J_\perp \gg J$) appears to be valid even at isotropic couplings $J_\perp = J$. For undoped spin-$\frac{1}{2}$ ladders, there is *no spin gap* when the number of chains is *odd*, and a *finite spin gap* when the number of chains is *even*. Doped ladders are not yet well understood.

In the next chapter we give a brief overview of real ladder-materials. The focus of chapter 3 are *undoped* systems. *Doped* systems correspond to the tJ and the Hubbard model, reviewed in chapter 4. The paper finishes with a summary.

Springer Proceedings in Physics, Volume 82
Computer Simulation Studies in Condensed-Matter Physics IX
Eds.: D. P. Landau, K.K. Mon, H. -B. Schüttler

2 Real ladder materials

$(VO)_2P_2O_7$ is a 2-chain material that has been known for some years. Most other real ladder materials are derived from high T_c superconducters. The series $SrCu_nO_{2n-1}$ exhibits $n = 2, 3, ...$ coupled chains, and there are $n = 4, 5, ...$ coupled chains in $(La_2CuO_4)_{n-2}La_2CuO_7$. Fig. 1 shows how these materials are structured. In (a) the two-dimensional precursor is shown, with Cu atoms at solid dots and oxygen atoms at intersections of solid lines. Line defects like in (b) and (c) isolate finite ladders from each other. The coupling between copper atoms across the defect is weak because it involves a 90 degree angle at the oxygen and is frustrated. ¿From this structure it also follows that the ladder materials should have more or less isotropic couplings.

Experimentally these materials are rather difficult to produce. For a review of experimental findings see [2]. Most experiments have been limited to *undoped* materials. Recently, *doped* 2-chain materials have also been produced and analyzed [2].

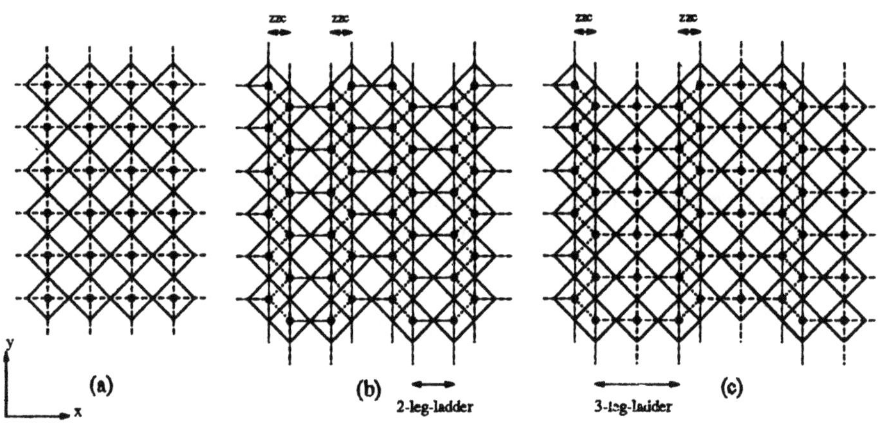

Figure 1: Schematic structure of the CuO planes in ladders derived from high T_c materials. The ladders are separated by linelike defects. (From [11]).

3 Undoped ladders: Heisenberg model

The undoped precursors of high T_c materials are well described by the spin-$\frac{1}{2}$ antiferromagnetic Heisenberg model. Thus this model should also be appropriate for ladder materials. Its Hamiltonian is

$$H = J \sum_{\parallel} \vec{S}_i \vec{S}_j \; + \; J_\perp \sum_{\perp} \vec{S}_i \vec{S}_j \; , \tag{1}$$

where J and $J_\perp$ are the couplings parallel and perpendicular to the chain direction, as denoted in fig. 2.

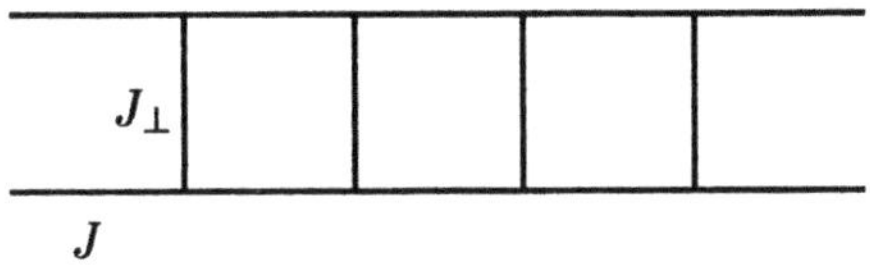

Figure 2: Schematic structure of a 2-chain ladder, with two couplings $J_\perp$ and J.

Both the single chain and the fully two-dimensional model on the square lattice are well understood. The single chain is exactly solvable by Bethe ansatz. It has no energy gap, and no long range order. The two-dimensional model has been extensively studied both by analytical methods (mapping to a nonlinear sigma model) and by numerical studies. There is now little doubt that it also has no energy gap, but long range order. Recent numerical studies of the 2d Heisenberg antiferromagnet [3], using the loop algorithm (see below), were able to make contact with chiral perturbation theory and to determine the free constants in the latter to very high precision.

The *dimensional crossover* from 1 to 2 dimensions in the Heisenberg model is expected to be *far from smooth*: Both the single chain and the two dimensional model are gapless. Yet ladders of *even* width are expected to have an energy gap, whereas those of odd width are again expected to be gapless. This expectation is based on the simple picture of largely independent rungs in the limit $J_\perp \gg J$, which is described in the next section. An open question is whether for very small $J_\perp$ there is qualitatively different behavior, i.e. a phase transition at finite $J_\perp$ from an ungapped to a gapped phase. ¿From existing studies (perturbative as well as numerical) there is no indication for such a phase transition.

The spin-gap behavior is analoguous [4] to that of single spin chains of spin S: when S is half-integer, the latter are known by the Lieb-Shultz-Mattis theorem [5] to be gapless. (A similar proof for odd-width ladders was recently given in [6].) When S is integer, the famous Haldane conjecture [7] predicts a spin gap of order $\exp(-\pi S)$. If the coupling $J_\perp$ between chains in spin ladders were ferromagnetic, then in the limit $(-J_\perp) \gg J$ the n-chain ladder could be mapped to a single chain of spin $S = \frac{1}{2}n$. However, since $J_\perp$ is actually antiferromagnetic, the relation is not so simple. A possible mapping from antiferromagnetic to ferromagnetic $J_\perp$ is described in section 3.2.

Coupled Heisenberg chains have been extensively studied in the past few years, both analytically and numerically. For an overview of this work, see [1]. In this paper, we shall focus on some selected recent studies.

3.1 Rung picture

The qualitative behavior of spin ladders can be well understood by studying the case $J_\perp \gg J$, as advocated by Dagotto, Rice, and collaborators. The resulting picture remains useful in the isotropic case, and even for $J_\perp < J$.

Let us explain the case of 2 coupled chains. When $J_\perp \gg J$, the *rungs* of the ladder become approximately independent. The groundstate then consists of spin singletts on the rungs ($k_y = 0$, where y is the direction of the rung). The first excited state is that with a single triplett ($k_y = \pi$) on one of the rungs, commonly called a "magnon". It behaves like a hard core boson. The dispersion relation is

$$\epsilon_k = J_\perp + J \cos k_x \quad , \tag{2}$$

and there is a finite gap

$$\Delta \approx J_\perp - J \quad . \tag{3}$$

The higher excited states then form a continuum of states. For $J_\perp \gg J$ the dispersion (2) of the three one-magnon branches does not overlap that of the two-magnon continuum. For $J_\perp = J$ the one-magnon branches are still largely separate from the continuum.

For three coupled chains the rung ground state at large $J_\perp/J$ will be a doublet, like in a single chain. These doublets are weakly coupled by J, leading to a gapless phase, similarly as for single chains. For a recent proof of gaplessness see [6].

3.2 Are the two-chain ladder and the $S = 1$ chain in the same phase ?

The fact that the spin gap of n coupled spin-$\frac{1}{2}$ chains behaves in qualitatively the same way as that of a single spin $S = \frac{1}{2}n$ chain would be very natural for *ferro*-magnetically coupled chains at $-J_\perp \gg J$, since then the spins on each rung would combine to an effective single spin of size $S = \frac{1}{2}n$.

Could it be that the antiferromagnetically coupled ladders ($J_\perp > 0$) and the ferromagentically coupled ones ($J_\perp < 0$) are in the same phase ? Then there should be a continuous path in parameter space from $J_\perp > 0$ to $J_\perp < 0$ which does not go through a phase transition.

For the case of 2 coupled chains, White has given strong arguments in a recent paper [8] that this is indeed the case. Note that the obvious path of just varying $J_\perp$ *does* pass through a transition, since at $J_\perp = 0$ the chains are decoupled and thus gapless, i.e. in a different universality class than the two coupled chains.

White takes the case $J_\perp = J$ and introduces an additional coupling of strength J_2 between sites $(i, 1)$ and $(i+1, 2)$ $(i = 1, 2, 3, ...)$. When $J_2 = 0$, the original antiferromagnetically coupled chains are recovered. When $J_2 \ll 0$, then sites $(i, 1)$ and $(i+1, 2)$ are strongly coupled ferromagnetically, so that the spins there will effectively form a single spin of size $S = \frac{1}{2} \cdot 2$.

In a numerical investigation of this model with the density matrix renormalization group, White varied the strength of J_2 while measuring the spin gap. The gap varied only slightly, giving no indication of a phase transition. Thus the gapped phases of two ferromagnetically and two antiferromagnetically coupled chains may indeed be the same. Note however that compared to two antiferromagnetically coupled chains, the modified chain with $-J_2 \gg J$ possesses two additional spins at the ends of the chain, which are coupled only with $J_\perp$ and J.

3.3 Numerical results: DMRG

In DMRG, the lowest states (i.e. zero temperature) of a finite system are iteratively computed while increasing the system size. Isotropically coupled chains were studied by White, Noack, and Scalapino [9], for up to $n = 4$ coupled chains and chain lengths of 24 at $n = 4$. By finite size analysis of the results they do indeed find a spin gap consistent with zero for $n = 1$ and $n = 3$. For the even width ladders they obtain gaps

$$\Delta = 0.504\,J \text{ for 2 chains,} \quad \Delta = 0.190\,J \text{ for 4 chains,} \tag{4}$$

directly from the energy differences of the lowest lying states. (The ladder length at $n = 4$ is rather small, so that the extrapolation to $L = \infty$ might be biased.)

3.4 Numerical results: Quantum Monte Carlo

Detailed results for the thermodynamics of large ladder systems up to 6 chains wide and 100 sites long have been obtained at low temperatures by Frischmuth, Ammon, and Troyer [10, 11]. We shall describe this work in some detail here. It has recently been confirmed and extended [12]. Other Quantum Monte Carlo studies have been performed by Hatano et al. [13] and by Sandvik et al. [14], the latter including dynamics.

The Quantum Monte Carlo studies [10, 12] were made possible by means of the "loop algorithm" [15], which shall be described briefly now. It is a quantum Monte Carlo method in the worldline formulation that employs *nonlocal* changes of worldline configurations and thereby overcomes autocorrelations. Because the Hamiltonian preserves total S_z and couples only locally, all worldlines and all possible *changes* of worldlines (present $\leftrightarrow$ not present, i.e. spin up $\leftrightarrow$ spin down) are located on closed loops, as exemplified in fig. 3. These loops are constructed by stochastic *local* choices of direction, so that no acceptance steps with global weights occur. As a consequence the method is *ergodic*, for example in total magnetization and in worldline winding number.

A further advantage is the existence of "Improved Estimators", which are measurements that average implicitely over 2^{n_l} worldline configurations, where n_l is the number of loops constructed during an update. Thus these improved estimators can have drastically reduced statistical errors. For example, the uniform spin susceptibility can be written

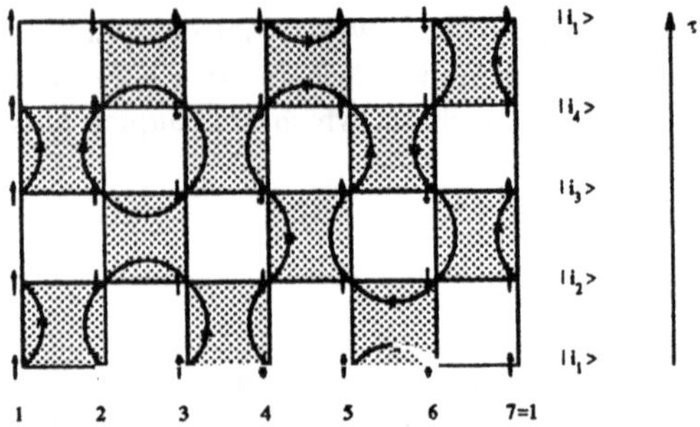

Figure 3: Example of a loop-configuration in the Heisenberg model. The loops connect sites where spins will be flipped together. ([11]).

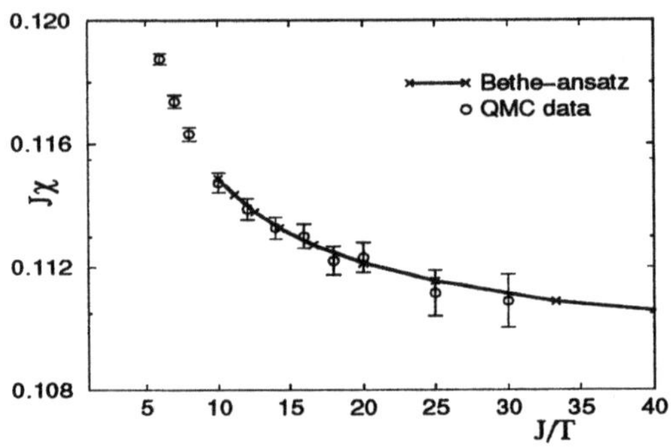

Figure 4: Comparison of QMC with exact results from Bethe ansatz, for a single isotropic chain. (Adapted from [11].)

$$\chi \sim \underbrace{\langle \sum_{ij} S_i^z S_j^z \rangle}_{noisy} = \langle \left(\sum_{\text{loops } \alpha} \sum_{\text{sites in } \alpha} S^z \right)^2 \rangle = \tfrac{1}{4} \underbrace{\langle \sum_{\text{loops } \alpha} W_\alpha^2 \rangle}_{stable} , \quad (5)$$

where W_α is the 'winding number' of loop number α in temporal direction.

The elimination of autocorrelations and the use of improved estimators together resulted in a gain of computer time in [10] of more than 5 orders of magnitude, compared to the standard local worldline QMC method. (For a different good Monte Carlo method, see [14]). This made it possible to go to very large systems of up to 6 chains of length 100, at very low temperatures, down to $T/J = 0.02$ ($\beta J = 50$).

Fig. 4 compares the QMC results for the susceptibility of a single chain with the exact Bethe ansatz result showing that the Monte Carlo calculations are indeed reliable.

The ground state energies of isotropic ladders are given in table 1. They are obtained by extrapolating the Monte Carlo results to zero temperature. One

Table 1: Ground state energies per site for ladders of different widths. ([10].)

number of legs	E	reference value
1	-0.4432(1)	-0.44315... (Bethe)
2	-0.5780(2)	-0.578
3	-0.6006(3)	-
4	-0.6187(3)	-
5	-0.6278(4)	-
6	-0.635(1)	-
2D lattice		-0.6693(1)

sees clearly that the ground state energy is a *smooth* function of the ladder width, in contrast to the spin gap.

The central result is obtained from the spin susceptibility, shown in fig. 5. Both subfigures depict the same data, plotted in (a) against temperature, and in (b) against inverse temperature. The behavior at large temperatures is well described by series expansion. One clearly sees the qualitative difference between ladders of even and of odd width. When there is a gap, then in the limit $T \to 0$ the susceptibility should behave like

$$\chi \stackrel{T\to 0}{\longrightarrow} \frac{e^{-\Delta/T}}{T^{\frac{1}{2}}}. \tag{6}$$

(Here a plausible form [16] for the dispersion relation has been assumed to obtain the power law exponent). On the log-log-plot of fig. 5(b), eq. (6) results in straight lines at large $J\beta$, which are clearly visible in the figure. A fit to their slope gives the gap

$$\Delta = \begin{cases} 0.51(1) & ,2 \text{ legs} \\ 0.17(1) & ,4 \text{ legs} \\ 0.05(1) & ,6 \text{ legs} \end{cases} \tag{7}$$

(with very conservative error bars).

Without a gap, one expects

$$\chi \stackrel{T\to 0}{\longrightarrow} \text{const} \tag{8}$$

In fig. 5(b), the ladders of odd width still show a slight slope of approximately $0.005J$, which is of the expected magnitude of a finite size effect for ladders of length 100. We can thus conclude that the ladders of odd width are indeed gapless. Note that the asymptotic $T \to 0$ behavior is reached, as expected, at progressively smaller temperatures of order Δ as the gap decreases.

Further analysis of the ladder systems [11] shows that those of width 2 can be described by a single-magnon picture (as in the rung-picture), even at $J_\perp = J$. Ladders of width 4 can be mapped to two 2-leg-ladders, with one bonding and one anti-bonding branch. The gapless 3 and 5-leg ladders can

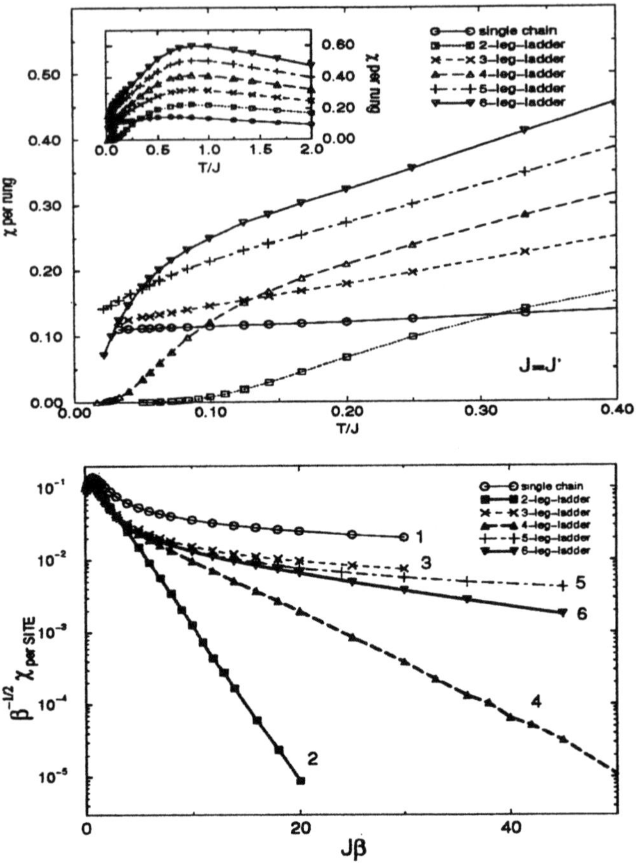

Figure 5: Uniform susceptibility for ladders of different width. Top: χ against temperature ([10]). Bottom: The same data, with $T^{\frac{1}{2}}\chi$ plotted logarithmically against inverse temperature.

be mapped approximately to a single chain at an effective coupling J_{eff} at low temperatures. The susceptibility appears to depend *smoothly* on $J_\perp/J$ for $J_\perp > J$. The question whether there is a phase transition at small $J_\perp$ was not addressed in [10, 11].

Comparison of the susceptibility to experimental data gives slightly varying results for J. For two-leg ladders, the experimentally measured value $\Delta \approx 420K$ gives $J \approx 840K$ from eq. (7). For three-leg ladders a fit to data from $Sr_2Cu_3O_5$ results in $J \approx 1100K$. The differences in J may be caused by the effects of residual interactions between different ladders.

In summary, the expected dependence of the gap on ladder width is indeed clearly seen, with gaps in ladders of even width, and gapless behavior in ladders of odd width. Note also that the gaps in eq. (7) are consistent with an exponential decrease with number n of chains, which would be similar to the Haldane behavior $\Delta \sim e^{-\pi S}$ for single chains of spin S. Eq. (7) does not look like the $1/n$ behavior which had been suggested before [9].

4 Doped ladders: tJ model

Doped ladders, i.e. the ladder versions of high temperature superconductors, are especially interesting. Unfortunately, they are very difficult to study both experimentally and theoretically. On the numerical side the results come mostly from exact diagonalization studies of rather small systems of size e.g. 10 × 2 [19, 20, 21, 22] The relevant models for doped ladders are the tJ and the Hubbard model.

Let us describe the basic issues for the case of the tJ model. Its Hamiltonian is

$$\begin{aligned} H = \; & -\sum_{<ij>,\sigma} t_{ij}(c^{+}_{i\sigma}c_{j\sigma} + h.c.) \quad (t_{ij} = t, t_{\perp}) \\ & +\sum_{<ij>} J_{ij}(\vec{S}_i\vec{S}_j - \tfrac{1}{4}n_i n_j) \quad (J_{ij} = J, J_{\perp}) \; . \end{aligned} \tag{9}$$

At each site there is either an electron of spin up or of spin down, or a hole. At half filling (nondoped case), the tJ model reduces to the Heisenberg model.

A *single* chain in the tJ model is known to be a so called Tomonaga-Luttinger liquid (see below), which has no spin gap, no charge gap, and spin-charge separation.

Experimentally, 2-chain ladders appear to have a spin gap even when doped. The most interesting theoretical issues are to understand this behavior and to see whether there is a tendency towards superconductivity upon doping.

4.1 Luttinger Liquids

Single chains in the tJ and in the Hubbard model are "Tomonaga-Luttinger Liquids" [17], exhibiting spin-charge separation. Two universality classes are important here. (Actually they are groups of universality classes, since they depend on a parameter K_ρ). In the *Tomonaga-Luttinger liquid (TL)*, both spin and charge are gapless. In the *Luther-Emery liquid (LE)*, only the charge mode is gapless. In both cases, correlations decay with a component that behaves like a power law in inverse distance $1/r$, with an exponent that depends on the single parameter K_ρ. These exponents are given in table 2. Note that at large distances, pairing correlations dominate when $K_\rho > 1$. (However, in a 1-dimensional system at finite temperature there cannot be true long range order. The ladder systems are also 1-dimensional because of their finite width.) Numerical data even on relatively small systems can give information about K_ρ by utilizing conformal field theory.

The case of two coupled Luttinger liquids has been investigated by Schulz [18]. The results are given as "2LL" in table 2. Now both spin and charge

Table 2: Correlation exponents. (For "2LL", the transverse momentum is π).

	TL	LE	2LL
$2k_f$ SDW	$1+K_\rho$	exponential	exponential
$2k_f$ CDW	$1+K_\rho$	K_ρ	exponential
Singlett pairing	$1+\frac{1}{K_\rho}$	$\frac{1}{K_\rho}$	
Triplett pairing	$1+\frac{1}{K_\rho}$	exponential	
d-wave pairing			$\frac{1}{K_\rho}$
$4k_f$ CDW	$4K_\rho$	$4K_\rho$	K_ρ

modes are massive, and again superconducting correlations are expected to dominate when $K_\rho > 1$.

4.2 tJ model: Rung picture

The rung picture, i.e. the case $J_\perp \gg J$, can also be used to understand coupled tJ-chains [21]. Let us consider the rung picture for 2 coupled chains. In the exact-diagonalization studies described later, the rung picture appears to survive to the case of isotropic coupling (with bound pairs becoming spatially more extended).

In case of half filling, we have the Heisenberg spin model discussed before. If to this we add a single hole, it will behave like a quasiparticle with both spin and charge. When we add 2 holes, they can either locate on the same rung, or on different rungs. In the ground state, they will be on a single rung to minimize the number of broken rungs of energy cost $J_\perp - t - t_\perp$. Their binding energy corresponds to breaking another rung, i.e.

$$E_{binding} \approx J_\perp - t - t_\perp \ . \tag{10}$$

For energies below $E_{binding}$ we should then expect hole pairs, behaving like hard core bosons with weak attractive interactions $\sim J$, and thus dominant pair correlations, which are "d-wave-like" because of momentum π in rung-direction.

Spin excitations are then gapped. They can either consist of broken pairs of holes with energy $E_{binding}$, or spin-magnons like in the Heisenberg model, with energy $\sim J_\perp$. Since $E_{binding} < J_\perp$, a jump in the gap can be expected between half filling and finite doping. Charge excitations correspond to coherent motion of bound hole pairs and should be gapless. Gapped spin and gapless charge modes are reminiscent of the Luther-Emery liquid.

4.3 tJ model: some results from exact diagonalization

A tentative phase diagram of two coupled tJ chains has been given by Poilblanc et al. [20], for the isotropic case $J_\perp = J$, $t_\perp = t$. At $J/t < 2$ and fillings between about $\frac{1}{2}$ and 1, they do indeed see a phase that resembles the prediction from the rung picture: there is a spin gap, no charge gap, and d-wave like hole pairing. At large $J/t > 4$ they see phase separation between holes and electrons for all fillings. Phase separation minimizes the antiferromagnetic energy. For $J/t < 2$ and filling below $\frac{1}{2}$, the system appears to behave like a Tomonaga-Luttinger liquid, with no gap. At very small filling and $2 < J/t < 4$ there is another phase, with a spin gap and electron pairs.

Interestingly, two coupled chains may behave very differently, depending on whether each *single* chain is integrable or non-integrable. Mila and Poilblanc [23] claim that at the integrable points $J/t = 0$ and $J/t = 2$, and only there, hopping between the chains is "coherent", i.e. it has a periodic component in its time-dependence. This is an area of intense current research.

4.4 Hubbard model

The Hubbard model is a somewhat more realistic description of the doped systems, since it allows two electrons of opposite spin to occupy the same site. Its Hamiltonian is

$$H = - \sum_{<ij>,\sigma} t_{ij}(c_{i,\sigma}^{+} c_j + h.c.) + U \sum_i n_{i\uparrow} n_{i\downarrow} \ . \tag{11}$$

For this model there exist both weak coupling renormalization group studies and numerical results [24, 25, 26, 27]. At half filling, the dynamical properties have also been investigated [28], with results consistent with a rung picture.

The main features of the phase diagram can most easily be understood from the $U = 0$ case [25], i.e. the free case. Then there are a bonding and an antibonding band, with dispersion

$$\epsilon_k = -(2t \cos k \ + \ t_\perp \cos k_\perp), \quad k_\perp = 0, \pi \ . \tag{12}$$

For $t_\perp < 2t$ the two bands overlap. When some interaction is switched on, there should be Umklapp scattering, producing a Luther-Emery like phase with a spin gap. For $t_\perp > 2t$ the bands are completely separate, and one expects a Tomonaga-Luttinger liquid, without any gap. Thus the simple phase diagram of fig. 6 is obtained.

Balents and Fisher [29] have studied two coupled Hubbard chains by weak coupling renormalization group for infinitesimally small positive U. The resulting phase diagram is shown in fig. 7. Its main structures are very similar to the $U = 0$ case in fig. 6. The phases are denoted as $CnSm$, meaning that there are n gap*less* charge modes and m gap*less* spin modes. The C1S1 phase (no gaps) is like a Tomonaga-Luttinger liquid. C2S2 is the analogue of a Fermi liquid. In the C1S0 phase, there is a spin gap but no charge gap, and d-wave like pairing. Overall the phase diagram exhibits a very rich structure. It is predicted [29] to become even richer as U becomes larger, and remains to be confirmed by independent means.

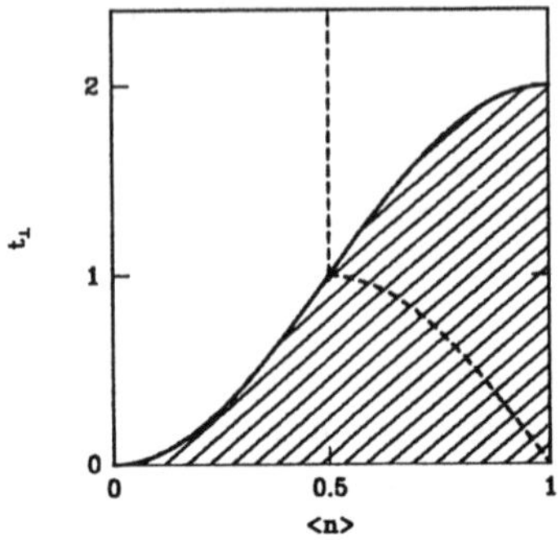

Figure 6: Phase structure for two coupled Hubbard chains at $U = 0$. In the shaded area, both the binding and the antibinding band are occupied. On the dashed line, the lower band is $\frac{1}{2}$ filled. ($t_\perp$ is given in units of t). ([25]).

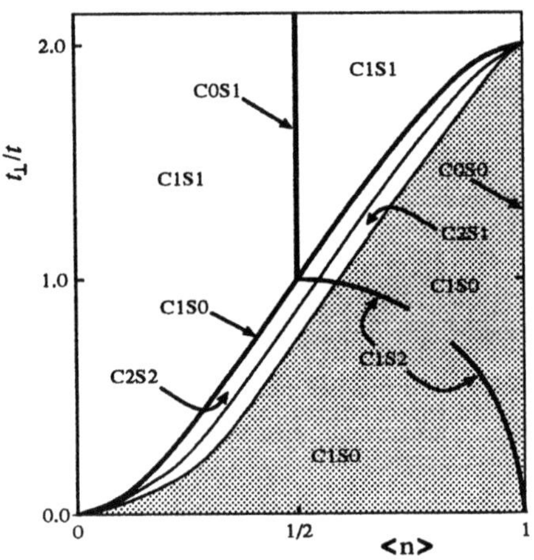

Figure 7: Phase diagram of two coupled Hubbard chains at infinitesimal U. ([29]).

Noack et al. [24, 25] investigated two coupled Hubbard ladders at *strong* coupling $U = 8$ by means of DMRG, on systems of size 2×8 up to 2×64. Surprisingly, at the couplings investigated they find behavior consistent with Balents and Fisher, specifically a C1S0 phase which is Luther-Emery like, showing a spin-gap and d-wave like singlett pairs of holes.

For *three* coupled chains, there is a weak coupling renormalization group calculation by Arrigoni [30], similar to that of Balents and Fisher. The resulting phase diagram, shown in fig. 8, again has a very rich structure of phases. It remains a challenge to verify this structure by other methods.

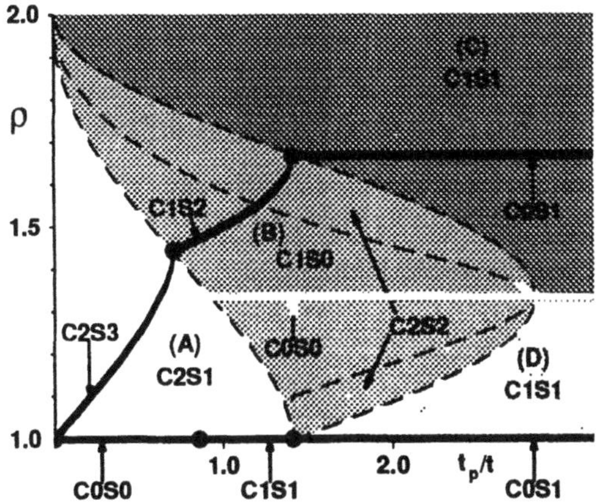

Figure 8: Phase diagram of three coupled Hubbard chains at infinitesimal U. ([30]).

5 Conclusions

Systems of coupled quantum chains, also called ladder systems, are related to the high T_c superconducting materials. They show a very rich and interesting structure, and allow to study the dimensional crossover from one to two dimensions. Much of their behavior can be understood even for the isotropic case in a simple rung picture obtained at large transverse coupling.

Undoped ladders can be described by the Heisenberg spin $\frac{1}{2}$ antiferromagnet, and are rather well understood. For an odd number of coupled chains, they have no spin gap; for an even number of coupled chains, they have a finite gap, which decreases with increasing number of chains. It can be related to the Haldane gap of single spin-S chains. Numerical studies of ladders, including Quantum Monte Carlo studies of up to 6 coupled chains at very low temperature, confirm the expectation and match well to experiment.

Doped ladders may be described by the tJ or the Hubbard model. They are far less well understood, both analytically and numerically. They possess a very rich phase diagram, which includes phases with d-wave like pairing that might evolve into superconducting phases as the number of chains goes to infinity. The phase diagrams proposed from weak coupling renormalization group remain to be explored by other methods.

Acknowledgements

This research was supported by BMBF grant # 05605WWA. I would like to thank B. Ammon, E. Arrigoni, B. Frischmuth, R. Noack, and M. Troyer for stimulating and enlightening discussions.

References

[1] E. Dagotto and M. Rice, Science **271**, G18 (1996), and references therein.

[2] Z. Hiroi and M. Takano, Nature **377**, 41 (1995), and references therein.

[3] U.-J. Wiese and H.-P. Ying, Z. Phys. **B93** (1994) 147;
B.B. Beard and U.-J. Wiese, cond-mat/9602164.

[4] G. Sierra, cond-mat/9512007.

[5] E. Lieb, T.D. Schultz, and D.C. Mattis, Ann. Phys. **16**, 407 (1961).

[6] For a proof, see A.G. Rojo, Phys. Rev. **B53**, 9172 (1996).

[7] F.D.M. Haldane, Phys. Lett. **A93**, 464 (1983), Phys. Rev. Lett. **50**, 1153 (1993).

[8] S.R. White, Phys. Rev. **B53**, 52 (1996).

[9] S.R. White, R.M. Noack, and D.J. Scalapino, Phys. Rev. Lett. **73**, 886 (1994).

[10] B. Frischmuth, B. Ammon, and M. Troyer, cond-mat/9601025.

[11] B. Frischmuth, thesis, and private communication.

[12] M. Greven, R.J. Birgeneau, and U.-J. Wiese, cond-mat/9605068.

[13] N. Hatano and Y. Nishiyama, J. Phys. **A28**, 3911 (1995), Y. Nishiyama, N. Hatano, and M. Suzuki, J. Phys. Soc. Jpn. **65**, 560 (1996).

[14] A.W. Sandvik, E. Dagotto, and D.J. Scalapino, Phys. Rev. **53**, R2934 (1996).

[15] H.G. Evertz, G. Lana, and M. Marcu, Phys. Rev. Lett. **70**, 875 (1993).

[16] M. Troyer, H. Tsunetsugu, and D. Würtz, Phys. Rev. **50**, 13515 (1994).

[17] See for example the reviews by J. Voit, Rep. Prog. Phys. **58**, 977 (1995), and H.J. Schulz, cond-mat/9503150.

[18] H.J. Schulz, Phys. Rev. **B53**, R2959 (1996).

[19] D. Poilblanc, H. Tsunetsugu, and T.M. Rice, Phys. Rev. **B50**, 6511 (1994).

[20] D. Poilblanc, D.J. Scalapino, and W. Hanke, Phys. Rev. **B52**, 6796 (1995).

[21] H. Tsunetsugu, M. Troyer, and T.M. Rice, Phys. Rev. **B49**, 16078 (1994); M. Troyer, H. Tsunetsugu, and T.M. Rice, Phys. Rev. **B53**, 251 (1996).

[22] C.A. Hayward, D. Poilblanc, R.M. Noack, D.J. Scalapino, and W. Hanke, Phys. Rev. Lett. **75**, 926 (1995), C.A. Hayward, D. Poilblanc, and D.J. Scalapino, Phys. Rev. **B53**, R8863 (1996), C.A. Hayward and D. Poilblanc, Phys. Rev. **B53**, 11721 (1996).

[23] F. Mila and D. Poilblanc, Phys. Rev. Lett. **76**, 287 (1996).

[24] R. Noack, S.R. White, and D.J. Scalapino, Phys. Rev. Lett. **73**, 882 (1996).

[25] R. Noack, S.R. White, and D.J. Scalapino, cond-mat/9601047.

[26] Y. Munehisa, cond-mat/9510039.

[27] Y. Asai, Phys. Rev. **B52**, 10390 (1995).

[28] H. Endres, R.M. Noack, W. Hanke, D. Poilblanc, and D.J. Scalapino, Phys. Rev. **B53**, 5530 (1996).

[29] L. Balents and M.P.A. Fisher, Phys. Rev. **B53**, 12133 (1996).

[30] E. Arrigoni, cond-mat/9509145.

Divide-and-Conquer Treatments of Electron Correlations

S. Liang

Department of Physics, 104 Davey Laboratory, The Pennsylvania State University, University Park, PA 16802, USA

Abstract Wilson's numerical renormalization group treatment of a correlated system requires the existence of an extended range of energy scales in the hamiltonian. The presence of such an energy scale hierarchy is often not evident. We argue here that the hierarchy of scales in the correlation energy can be brought about by a suitable transformation of the single particle orbitals. We formulate a variational procedure to select the optimal orbitals for a general hamiltonian. Our method is a natural generalization of the Hartree-Fock theory with systematic inclusion of correlation effects. It is therefore completely general applicable to real systems such as atoms and molecules. As a test calculation, the method is implemented for a exactly solvable model. Comparing to the exact spectrum, we demonstrated, for the low energy spectrum, a systematic improvement of our method over that of Hartree-Fock theory.

1 Introduction

As pointed out by Weisskoft[1] long time ago, the interaction among electrons can usually be ignored at low energy. The reason for this is that in quantum mechanics, the effects of the interaction are to scatter an electron from one state to another. At low energy the phase space for scattering becomes very small due to Pauli principle. Therefore, the single-particle based theories, such as the shell model for nuclear structure and the Landau Fermi liquid theory for electrons in a metal, work very well at low energy.

However, modern condensed matter physics is full of examples where electron correlations play a vital role: fractional quantum Hall effects[3], rare earth materials[4], and high T_c superconductors[2]. Progress in our theoretical understandings of these areas requires a reliable treatment for electron correlations. Also, how to effectively deal with the electron correlations is a central issue in quantum chemistry, which will also benefit from the new method we discuss.

Springer Proceedings in Physics, Volume 82
Computer Simulation Studies in Condensed-Matter Physics IX
Eds.: D. P. Landau, K.K. Mon, H. -B. Schüttler

The difficulty associated with treating correlations in a N-electron system is due to the exponential growth with N of the size of the basis set required to describe the many-body wave functions. Quantum Monte Carlo method[5], which is normally effective in dealing with large degrees of freedom, suffers from so called 'minus sign' problem[6] for electronic systems. It is difficult to reach low temperatures. In exact diagonalization method, the dimension of the Hilbert space grows exponentially with the number of electrons making it impossible to do calculation for a large number of electrons.

A potentially useful technique[7] is the renormalization group, in which one seeks to expand the low energy states in terms of a small number of many-body states by a divide-and-conquer technique. The numerical renormalization group (NRG) method[8] has been successfully applied to Kondo problems. As pointed out by Wilson[9], the success of the NRG is due to a special feature of the Kondo hamiltonian. The electrons' variables can be partitioned into those associated with shells around the impurity. Only the nearest neighboring shells interact. The key feature of the hamiltonian is that the strength of the interaction decreases exponentially away from the impurity. In this approach, one first ignores the weakly interacting shells and diagonalizes the hamiltonian of strongly interacting shells. One then keeps a fraction of the low energy states and allows them to mix with the next strongest interacting shell which has been omitted so far. This procedure is repeated many times. The energies of the discarded states at each step are always higher than the interactions left untreated because of the exponential decay of the interaction. Thus, the discarded states are not important as far as the low energy properties are concerned.

It is clear that the successful application of this paradigm requires the hamiltonian to exhibit a hierarchy of energy scales, which is unfortunately not apparent for a general hamiltonian. For example, in the real space renormalization group, the most efficient procedure of truncating states is by adding one site at a time. The interaction between the block and the added site is the hopping integral t or the exchange interaction J which remain fixed. This lack of energy scale is reason why the real space renormalization group performs poorly for lattice models[10]. By keeping the density matrix eigenstates, White[11] has recently developed a beautiful method which is proven to be very accurate for one-dimensional systems with short range interactions. However, it has been difficult to apply this method to higher dimensional system[12].

The question we would like to address is the following: for a given hamiltonian, is it possible to transform it so that the new hamiltonian exhibit a hierarchy of energy scales by suitably choosing single particle orbitals? We argue that the answer is affirmative—selecting suitable orbital is a promising way to explore. It is easy to convince oneself that the strength of a individual term can be changed in the transformed hamiltonian. In fact, for two dimensional Hubbard model, as will be discussed in details in Section 2, by dividing momentum variables in **k**-space into shells according their energies and by constructing a new set of orbitals in each shell in terms of angular harmonics around the Fermi surface it is possible for the hamiltonian written in the new orbitals to display an exponentially distributed hierarchy of energy scales[13]. While such a construction may be valid for a large class of hamiltonians, it is not optimal for numerical calculations. In Section 3, we formulate a general variational procedure which selects optimal orbitals for the NRG calculations.

Our method can be viewed as an improvement over the Hartree-Fock (HF) theory because of its systematic inclusion of correlation effects. Thus, our method is completely general, applicable to atoms and molecules. Our method becomes exact in the limit of weak interaction. Furthermore, since only the interactions left untreated in HF theory need to be included, the effective interactions are much reduced.

2 Explicit demonstration of energy scale hierarchy

We now demonstrate through an explicit construction that with suitable transformation of single particle variables, it is possible to identify a hierarchy of energy scales for a many-body hamiltonian. We believe our construction works for a large class of models but here we focus on Hubbard model in two dimensions. The Hubbard model can be written in the momentum space as:

$$H = \sum_{\mathbf{k}\sigma} \epsilon(\mathbf{k}) c^{\dagger}_{\mathbf{k}\sigma} c_{\mathbf{k}\sigma} + \frac{U}{N} \sum_{\mathbf{k},\mathbf{k}',\mathbf{q}} c^{\dagger}_{\mathbf{k}\uparrow} c^{\dagger}_{\mathbf{k}'\downarrow} c_{\mathbf{k}'+\mathbf{q}\downarrow} c_{\mathbf{k}-\mathbf{q}\uparrow}$$

where $c^{\dagger}_{\mathbf{k}\sigma}$ creates an electron with momentum $\mathbf{k}$ and spin σ and $\epsilon(\mathbf{k}) = 2t(\cos(k_x) + \cos(k_y))$ is the kinetic energy of a free electron tight-binding hamiltonian. We first divide states with good crystal momentum quantum number $\mathbf{k}_s$ into shells labeled by $s = \pm 1, \pm 2, ...$ according to their kinetic energies: if $\frac{|\epsilon(\mathbf{k}_s) - \epsilon_F|}{\epsilon_F} < \delta$ then the state $\mathbf{k}_s$ belongs to the shell with $s = \pm 1$; and if $\delta \Lambda^{|s|-1} < \frac{|\epsilon(\mathbf{k}_s) - \epsilon_F|}{\epsilon_F} < \delta \Lambda^{|s|}$ the state $\mathbf{k}_s$ belongs to the shell s with $|s| > 1$. The positive(negative) integers label the shells with $\epsilon(\mathbf{k}_s)$ above(below) the Fermi energy ϵ_F. Parameter δ specifies the width of the first shell and $\Lambda > 1$ controls the width of higher shells. Next, within each shell, we make a list of all states and rank them according to their angle $\theta_j = \tan^{-1}\frac{(\mathbf{k}_{sj})_x}{(\mathbf{k}_{sj})_y}$. We then introduce new variables by Fourier transformation $f_{sm\sigma} = \frac{1}{\sqrt{N_s}}\sum_j e^{\frac{2\pi i m}{N_s}j} c_{\mathbf{k}_j\sigma}$, where N_s is the total number of states $c_{\mathbf{k}_j}$ in the shell s. The Hubbard repulsive interaction (the U-term) can now be written in terms of $f_{sm\sigma}$:

$$\frac{U}{N}\sum_{s_1,m_1}\sum_{s_2,m_2}\sum_{s_3,m_3}\sum_{s_4,m_4} C(s_1m_1;s_2m_2;s_3m_3;s_4m_4) f^\dagger_{s_1m_1\uparrow} f^\dagger_{s_2m_2\downarrow} f_{s_3m_3\downarrow} f_{s_4m_4\uparrow}$$

$$C(\{s_i,m_i\}) = \frac{1}{\sqrt{N_{s_1}N_{s_2}N_{s_3}N_{s_4}}}\sum_{j_1j_2j_3j_4} e^{2\pi i\left(\frac{m_1j_1}{N_{s1}}+\frac{m_2j_2}{N_{s2}}-\frac{m_3j_3}{N_{s3}}-\frac{m_4j_4}{N_{s4}}\right)} \quad (1)$$

The sum in the last equation is constrained by the conservation law $\mathbf{k}_{j_1} + \mathbf{k}_{j_2} = \mathbf{k}_{j_3} + \mathbf{k}_{j_4}$. When m_i are not all the same, the scattering is effectively zero because the destructive interference[15]. The scattering terms are largest when all $m_i = 0$ where all the terms add in phase and the sums over all momenta are only limited by the conservation law. For the remaining terms with $m_i = m$, the scattering terms decay exponentially with m. Fig.1 shows the result of a calculation of $\frac{C(s_i,m_i=m)}{C(s_i,m_i=0)}$ (labeled "scattering" in the figure) for a 30×30 lattice with 6 momentum shells. Clearly, the interaction strength decays exponentially with m.

Unlike the Hubbard model in $\mathbf{k}$-space, which creates particle-hole pairs with equal weight, the Hubbard interaction written in f's has a hierarchy of energy scales and is therefore a better basis for renormalization group calculation. We can for example keep $f_{sm\sigma}$ for only $m < M$. Since the scattering is the strongest in the small m channels, getting rid of $f_{sm\sigma}$ at large m has a smaller effect on the ground state.

We now argue that the new variables have the right features. If the U-term is treated by perturbation theory in the $\mathbf{k}$-space and one asks which states are more important. Then certainly the important states will be particle-hole excitations near the Fermi surface because they have small excitation energy. Whereas the particle-hole excitations deep inside and outside of Fermi surface will be less important. However, because Pauli exclusion principle, the low energy particle-hole excitations near the fermi surface have vanishingly small phase space. Since there are great many more particle-hole pairs away from the Fermi surface, their combined effect could be large. In the new f-variable, s-waves will be treated first because they are most strongly interacting. The s-wave variables are the mixtures of the original $\mathbf{k}$-space variables with increasingly smaller fraction coming from degrees of freedom for shell further away from the

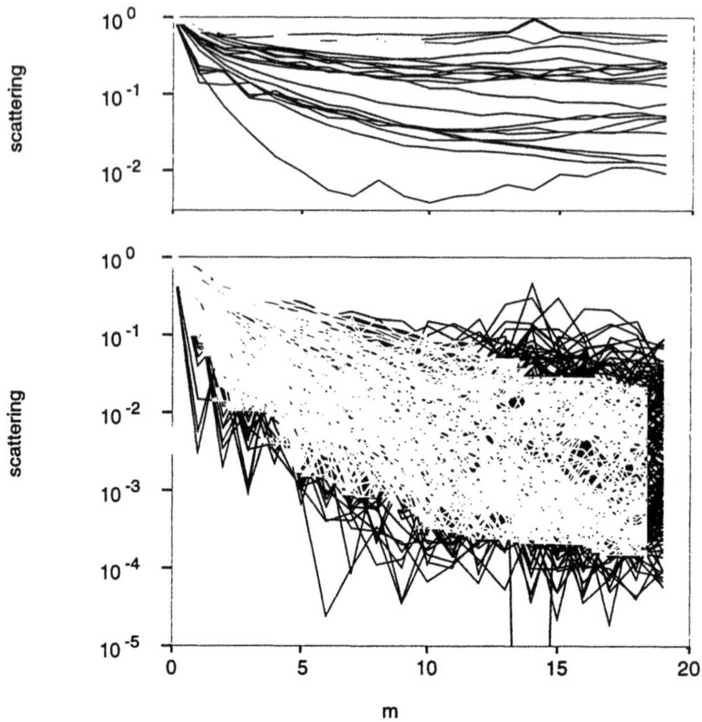

Figure 1: The coefficients $\frac{C(s_i, m_i = m)}{C(s_i, m_i = 0)}$ of Eq(1) (labeled "scattering" in the figure) for a 30×30 lattice with 6 momentum shells with $\epsilon_F = -1.3t$ and separation energies between shells at $|\epsilon - \epsilon_F| = 0.25t, t$. Each curve represents one of 6^4 configurations of (s_1, s_2, s_3, s_4). Only the terms with $m < 20$ are shown. The upper panel, which shows a slower decay, is for terms of the form $n_{sm\uparrow} n_{s'm\downarrow}$(with $n_{sm\sigma} = f^\dagger_{sm\sigma} f_{sm\sigma}$) that involve only the density and therefore no scattering(they contribute to diagonal energy only). The lower panel is for all the rest.

Fermi surface. This fraction is adjusted by the parameter Λ. Since we expect the momentum distribution $n(\mathbf{k})$ to change rapidly only near the Fermi surface, therefore more variables are needed there. By adjusting the parameter Λ, we achieve the right mixture.

In summary, we have given a heuristic argument demonstrating that with a suitable choice of the single-particle orbitals, for the two dimensional Hubbard model, the interaction energy exhibits a hierarchy of energy scales. This opens up a new possibility for studying strongly correlated electron systems using numerical renormalization group.

3 General scheme for identifying optimal orbitals

The explicit construction in the last section demonstrates the possibility of manipulating the single particle basis set in order to derive a hierarchy of energy scales essential for a renormalization group treatment. What is the best way to identify the optimal single particle basis set? This is the question we attempt to answer in this section. Our method is based on a generalization of Hartree-Fock method with systematic inclusion of electron correlations. The first step of our RG procedure is similar to the multiconfigurational self-consistent field method in quantum chemistry. Our method is therefore very general.

We illustrate our method using a one-dimensional spinless fermion model with nearest neighbor repulsion, described by the following hamiltonian:

$$H = -t \sum_i (c_i^\dagger c_{i+1} + c_{i+1}^\dagger c_i) + V \sum_i n_i n_{i+1} \tag{2}$$

where $c_i^\dagger$ creates a fermion on site i; $n_i = c_i^\dagger c_i$ is the density operator; and periodic boundary conditions are used. The hamiltonian can be solved exactly by Bethe Ansatz[16]. We compare the low energy spectrum calculated by our method with the exact results.

We begin by reviewing HF theory where orbitals are transformed to take into account the average interactions. The trial wave function, $|\psi_{HF}\rangle$, for the M-particle ground state of Eq. (2) is a Slater determinant with M orbitals filled. The M orthogonal single particle levels are determined by minimizing $\langle\psi_{HF}|H|\psi_{HF}\rangle$. The transformed orbitals are the eigenstates of the mean-field hamiltonian for the problem. By including particle-hole excitations from the HF ground state, we generate a complete set of many-body states. However, this complete set of states does not diagonalize the hamiltonian because the matrix element $\langle\psi_2|H|\psi_{HF}\rangle$ is in general non-zero, where $|\psi_2\rangle$ differs from $|\psi_{HF}\rangle$ by a two-particle-hole excitation. These off-diagonal matrix elements that remain untreated in the HF theory represent the effects of correlation. Our goal is to include in the fluctuating part of the wave function in a systematic fashion.

As indicated earlier, our choice of orbitals is motivated by the desire to find a hierarchy of energy scales in the correlation energy. We want to find the best orbitals so that when the Hilbert space is pruned, the low energy spectrum of the hamiltonian is not much affected. In our theory, the orbitals chosen will therefore differ from those in HF theory.

Let us take as an example a half-filled band of spinless fermions $N = 2M$, where N is the total number of independent orbitals and M is the number of particles. The new single particle orbitals are defined by $f_j =$

$\sum_i U_{ji}c_i$, where the transformation matrix is real and orthogonal: $U^T U = I$. In the first step of RG, we constrain the problem by limiting the scope of the fluctuation in the wave function.

$$|1_{RG}\rangle = \sum_{\alpha_1} \psi_{\alpha_1}|\alpha_1\rangle \tag{3}$$

$$|\alpha_1\rangle = \prod_{\substack{i=1 \\ M-M_1<k_i\leq M+M_1}}^{M_1} f_{k_i}^{\dagger}|1_{core}\rangle \tag{4}$$

In this trial wave function, $M - M_1$ 'core orbitals' are filled by $M - M_1$ particles: $|1_{core}\rangle = \prod_{p=1}^{M-M_1} f_p^{\dagger}|0\rangle$($|0\rangle$ is the vacuum state with no particle). The remaining M_1 particles are allowed to fluctuate among $2M_1$ 'valence orbitals'. There are $M - M_1$ orbitals that are completely empty. The trial wave function in Eq. (4) is expanded over $\binom{2M_1}{M_1} = \frac{(2M_1)!}{M_1!M_1!}$ Slater determinants.

Since $M_1 < M$, we have a Hilbert space which is much smaller than the Hilbert space of the full problem. We minimize $\langle 1_{RG}|H|1_{RG}\rangle/\langle 1_{RG}|1_{RG}\rangle$. Our variational parameters are the coefficients ψ_{α_1} and also, more importantly, the $M + M_1$ orthogonal orbitals specified by U_{ki}. The optimization process is divided into two steps. First, with orbitals fixed, the best coefficients ψ_{α_1} are chosen to be the ground state of the hamiltonian matrix $\langle\alpha_1'|H|\alpha_1\rangle$ in the reduced Hilbert space using the Lanczos method. Second, with the coefficients ψ_{α_1} fixed, the expectation energy is optimized with respect to U_{ki} using well established techniques of constrained optimization[17]. The details will be given elsewhere[18]. These two steps are repeated until the energy converges. Finally, we diagonalized the hamiltonian matrix $\langle\alpha_1'|H|\alpha_1\rangle$. The high energy eigenstates will be discarded later. It is sufficient to note now that the energy eigenstates serve equally well as a basis of the $\binom{2M_1}{M_1}$-dimensional space.

Comparing the HF ground state $|\psi_{HF}\rangle$ with the ground state trial wave function $|1_{RG}\rangle$, the optimally chosen valence orbitals will be such that the largest off-diagonal matrix elements ignored in the HF theory will be included in $|1_{RG}\rangle$. For large interaction, the energy splitting between the $\binom{2M_1}{M_1}$ eigenstates of $\langle\alpha_1'|H|\alpha_1\rangle$ will be determined by the largest off-diagonal energy, not by the lowest HF excitation energy. The low energy spectrum calculated in this step will not be correct. However, the large energy splitting makes the RG state truncation safe. This step is similar in spirit to the multi-configurational self-consistent field method.[19]

We now turn to the second step of RG: the Hilbert space in which the wave function is allowed to fluctuate is expanded. We allow one of the previously completely filled core orbitals(created by $f_{2core}^{\dagger}$) to become partially emptied. Similarly, we allow one of the previously empty orbitals

($f^\dagger_{2empty}$) to become partially filled. Conceptually, we write down a trial wave function similar to $|1_{RG}\rangle$:

$$|2_{RG}\rangle = \sum_{\alpha_2} \psi_{\alpha_2} |\alpha_2\rangle \tag{5}$$

$$|\alpha_2\rangle = \prod_{\substack{i=1 \\ M-M_2<k_i\leq M+M_2}}^{M_2} f^\dagger_{k_i} |2_{core}\rangle \tag{6}$$

with $|2_{core}\rangle = f_{2core}|1_{core}\rangle$ differing in that there are now $M_2 = M_1 + 1$ valence electrons fluctuating among $2M_2$ orbitals. The $2M_2$ valence orbitals are composed of the $2M_1$ valence orbitals from the first step, plus the ones created by $f^\dagger_{2core}$ and $f^\dagger_{2empty}$.

In the actual calculation, the RG approximation made in going from the first step to the second step involves pruning the $\binom{2M_1}{M_1}$-dimensional Hilbert space by discarding high energy eigenstates. The basis states $|\alpha_2\rangle$ involve direct product of the retained states from the first RG step and the orbitals created by $f^\dagger_{2core}$ and $f^\dagger_{2empty}$.

Since any of the core orbitals can become unfilled, we choose the partially unfilled orbital, $f^\dagger_{2core}$, to be a linear combination of all of the core levels in the first RG step in order to minimize the energy of the wave function $|2_{RG}\rangle$. We similarly define $f^\dagger_{2empty}$. By optimizing the new fluctuating orbitals created by $f^\dagger_{2core}$ and by $f^\dagger_{2empty}$, we ensure that the biggest correlation energy left untreated in the first step will be included in this second step. We apply this procedure repeatedly until the particles are allowed to fluctuate among all the orbitals.

One of important philosophy of renormalization group is to divide a big problem into smaller pieces. An iterative technique is then used to solve the series of small problems. In order for us to do this, additional states in various Hilbert space sectors has to be saved. Let $\mathcal{V}^n_m$ denotes the many body Hilbert space of m particles fluctuating among n valence orbitals. So that the dimensions of the Hilbert space $\mathcal{V}^n_m$ is $\binom{n}{m}$. The Hilbert space $\mathcal{R}^{N-n}_{M-m}$ similarly contains all the remaining $M-m$ frozen particles fluctuating among $N-n$ filled and empty orbitals. The full Hilbert space of M particles and N orbitals may be expressed as

$$\mathcal{W} = \sum_m \mathcal{V}^n_m \otimes \mathcal{R}^{N-n}_{M-m}$$

where the sum denotes the direct sum and $\otimes$ denotes the direct product. As we have discussed, in our treatment, $\mathcal{R}^{N-n}_{M-m}$ is always replaced by a single core state. This is true for all RG steps. At each RG step n is taken to be $2M_i$ and m is about M_i. However several sectors of $\mathcal{V}_m^{2M_i}$ with different particle quantum numbers m around M_i has to be saved.

To see this, let $\mathcal{E}_m^1$ denotes the Hilbert space created by $f_{empty}^\dagger$ with $m = 0, 1$ particle. And we similarly define $\mathcal{C}_m^1$ as the Hilbert space of the core level $f_{core}^\dagger$. The Hilbert space at a latter stage is built upon the previously ones. Since the particle number is conserved, the new Hilbert space is obtained using

$$\begin{aligned} \mathcal{V'}_{m+1}^{n+2} &= \mathcal{V}_{m+1}^n \otimes \mathcal{E}_0^1 \otimes \mathcal{C}_0^1 \oplus \mathcal{V}_m^n \otimes \mathcal{E}_1^1 \otimes \mathcal{C}_0^1 \oplus \\ & \qquad \mathcal{V}_m^n \otimes \mathcal{E}_0^1 \otimes \mathcal{C}_1^1 \oplus \mathcal{V}_{m-1}^n \otimes \mathcal{E}_1^1 \otimes \mathcal{C}_1^1 \end{aligned} \tag{7}$$

where the primed quantities on the left hand side of the equation denotes the new Hilbert space after one iteration and the unprimed quantities on the right hand side of the equation denotes the old Hilbert space before the iteration. The new Hilbert space after one iteration is built upon the old one.

We now discuss the procedure for state truncation, the Hilbert space of the valence electrons are divided into two parts:

$$\mathcal{V}_m^n = \mathcal{L}_{m,n} \oplus \mathcal{U}_{m,n} \tag{8}$$

where $\mathcal{L}_{m,n}$ and $\mathcal{U}_{m,n}$ are the lower and upper parts of energy eigenstates of the Hamiltonian in the restricted Hilbert space of $\mathcal{V}_m^n$. The division into the upper and the lower part is determined by keeping a pre-prescribed number of states in $\mathcal{L}_{m,n}$. Sectors $\mathcal{V}_{m\pm1}^n$ are similarly divided. When Eq.(8) is substituted into Eq.(7), the part associated with the upper part of the energy spectrum $\mathcal{U}_{m,n}$ will be discarded.

As we have discussed, once the valence orbitals are determined and the eigenstates with M_l particles fluctuating among $2M_l$ orbitals, $\mathcal{V}_{M_l}^{2M_l}$, are obtained, we also need to save states in sectors $\mathcal{V}_{M_l\pm1}^{2M_l}$, $\mathcal{V}_{M_l\pm2}^{2M_l}$ and so on. We used the following criterion to truncate the spectra for these sectors: let E_c be the cutoff energy in the sector $\mathcal{V}_{M_l}^{2M_l}$ which is obtained by counting up in energy until a prescribed number of states(usually about 100) is reached. The same $2M_l$ orbitals is used in $\mathcal{V}_{M_l+1}^{2M_l}$ as the ones determined in the main sector $\mathcal{V}_{M_l}^{2M_l}$. The division $\mathcal{V}_{M_l+1}^{2M_l} = \mathcal{L}_{M_l+1,2M_l} \oplus \mathcal{U}_{M_l+1,2M_l}$ is determined by the cutoff energy E_c. To keep the total particle number the same we empty one of the core orbitals which is chosen to minimize the ground state of the (M_l+1)-particle sector. Similarly, for the (M_l-1)-particle sector we place the extra particle in the empty orbital chosen to minimize the energy.

Because of the orbital selection procedure, the RG truncation step will not affect significantly the low energy properties. This is because the

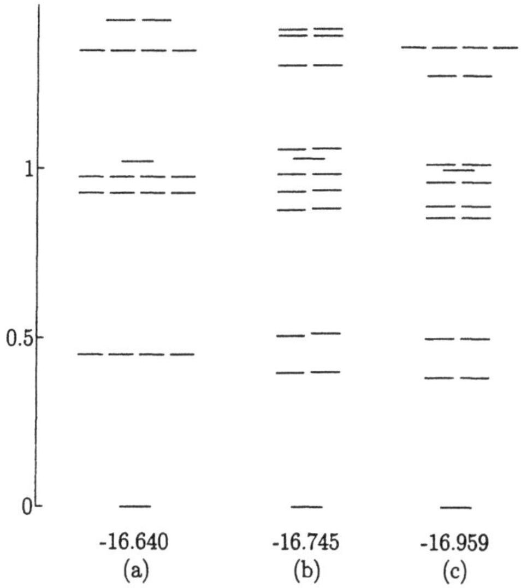

Figure 2: A comparison of the low energy spectra of (a) the Hartree-Fock theory, (b) adaptive RG calculation, and (c) the exact result for the hamiltonian Eq.(2) with 13 particles, 31 sites, and $V = t$. The number at the bottom of each spectrum is the ground state energy within each method. The lines represent the excitation energies of the 20 lowest excited states within each method(all energies are in the unit of t). In the RG calculation, we keep 100 states in the main sector. The total dimension of the Hilbert space is $\binom{31}{13} \simeq 2 \times 10^8$. The energy levels in the RG calculation are never exactly degenerate. For clarity of presentation, energy levels are drawn on multiple columns when their difference is smaller than $0.005t$.

energy splitting among the states in the first step is determined, when the interaction is large, by the biggest correlation energy, *i.e.* the off-diagonal matrix elements in HF theory discussed earlier. Further the energy splitting in the second step is caused by the remaining correlation energy not treated in the first step. Because of the way orbitals are chosen, the splitting in the second step will be smaller than that in the first step. Therefore there will not be a big change in the ordering of the energy levels and the truncated states remains at high energy even when further fluctuations are included in the latter steps. Therefore the effects of correlation energy in the latter RG steps will be smaller, in general. In another words, the states discarded earlier will not become low energy state when greater fluctuation is considered.

Typical results of the RG calculation are shown in Figs.2-4, in which the low energy spectra obtained by HF and our adaptive RG are compared to the exact results from Bethe Ansatz. In the case of Fig.2, the system has 13 particles moving on a 31-site ring with $V = t$. The dimen-

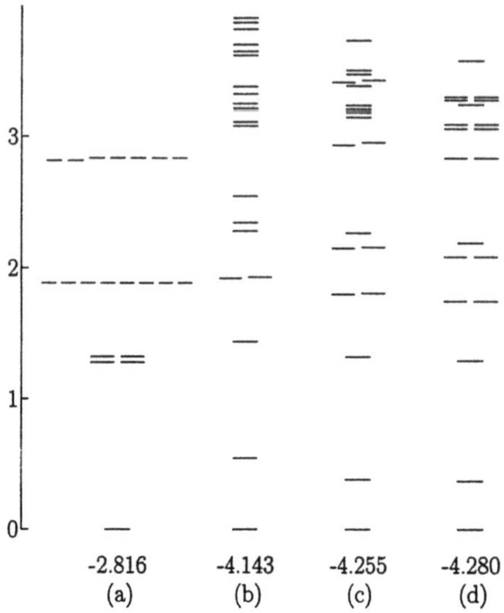

Figure 3: A comparison similar to Fig.2 for a half filled band(7 particles and 14 sites) at $V = 3t$. The system is in the Mott insulating phase and the correlation is therefore stronger. The energies of the lowest 20 states are plotted for (a) the Hartree-Fock theory; (b) adaptive RG with 100 states kept; (c) adaptive RG with 200 states kept; and (d) the exact result. By increasing the number of states retained in the RG calculations, the energies converge rapidly to the exact results.

sion of the Hilbert space is $\binom{31}{13} \simeq 2\times 10^8$. In the adaptive RG calculation, we keep 100 states each step. We start the RG calculation with 4 particles fluctuating among 13 valence orbitals. We then perform 9 RG steps in obtaining the spectrum in Fig.2. The HF excitation energies are computed directly from the HF wave function without using Koopmans' theorem. The adaptive RG spectrum represents a substantial improvement over the HF results. Comparing to the exact result, the errors in energy for a few lowest states is typically a few percent. This precision is comparable to that in the Kondo calculation of Wilson. However, unlike the Kondo hamiltonian, our starting hamiltonian does not have an apparent hierarchy of energy scales. For $V = 2t$(spectrum not shown), the error is slightly larger, but is always smaller than ten percent when 100 states are kept.

Fig.3 shows a comparison of the spectra for 7 particles and 14 sites at $V = 3t$, which for the infinite systems is in the Mott phase (from the exact solution, $V_c = 2t$) with a charge density wave. We cannot obtain the spectrum numerically from the Bethe Ansatz, so the exact result comes

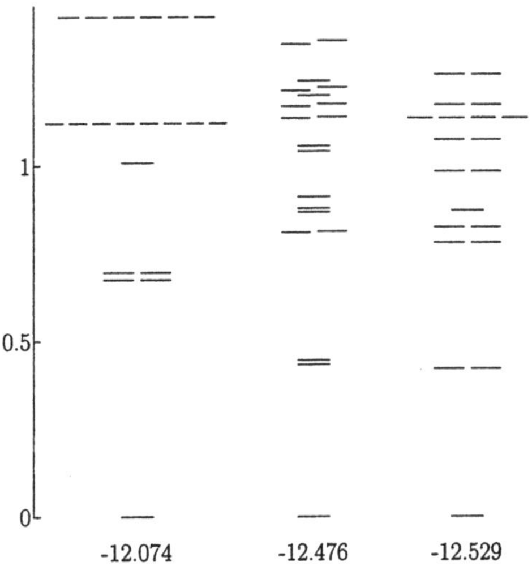

Figure 4: A comparison of the low energy spectra similar to Fig.3 but for attractive interaction V: (a) the Hartree-Fock theory, (b) adaptive RG calculation, and (c) the exact result. There are 7 particles, 14 sites, and $V = -1.5t$. The lines represent the excitation energies of the 20 lowest excited states within each method(all energies are in the unit of t). In the RG calculation, we keep 100 states in the main sector.

from diagonalizing the 3432×3432 hamiltonian matrix directly. We see that the accuracy increases rapidly as we increase the states kept from 100 to 200. Fig.4 shows the low energy spectra for attractive interaction $V = -1.5t$. Again, the renormalization group calculation is superior to the HF theory.

When the number of particles is odd, the exact spectrum of Eq.(2) is obtained by solving a set of non-linear equations, Eq.(18) of reference[16]. When the number of particles is even, the exact solution corresponds to the situation with one negative bond (π flux piecing through the ring). The spectrum is obtained by recursively making low energy particle-hole excitations upon the ground state configuration I_j in the Eq.(18) of reference[16].

Fig.5 shows the lowest 70 energy eigenstates in successive steps of renormalization group calculation. In the initial step, the average energy spacing between the levels is large. As the RG calculation progresses, more states appear at low energy. This observation is consistent with the notion that the core and empty states selected in the initial RG steps do not necessarily come from near the Fermi surface in the initial RG steps. Rather, they are selected because of their large correlation energy. In

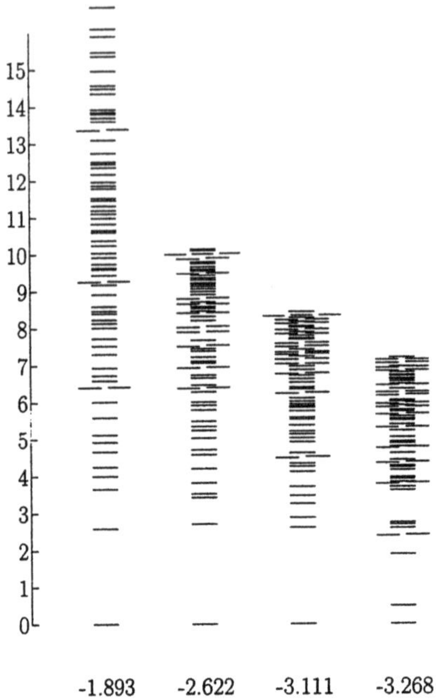

Figure 5: The low energy spectra of successive stages of the adaptive renormalization group calculation are shown with the earliest stage on the left. Notice the decreasing average energy spacing toward the latter stage of the RG calculation. The low energy spectrum is quantitatively correct only after several RG steps, and in this case in the last step. The calculation is done for $N = 14, M = 7$ and $V = 4$ with $M_1 = 4$ and total states kept 150.

Fig.5, there is also a graduate reduction of overall energy scale so that the energies are increasingly smaller as the RG procedure progresses.

As expected, the ground state energy is not precisely calculated because of the approximation involving large energy made to the deep core level excitations in the initial RG steps. However, as argued by Wilson[9] we expect the excitation energy to be better calculated in the RG approach. We expect our method to be applicable to systems in higher dimensions, which opens a new route to explore the novel physics of the high T_c superconductors[20, 21]. This is because we have not used any special structure associated with the one-dimensional lattice. The orbitals we choose are organized by the hierarchy of correlation energies, not by the one-dimensional connectivity that is used in the real space RG approach.

The effective interactions are much reduced in our method because the effects of interactions, on average, have been taken into account by the appropriately chosen core orbitals. The point is that although the core level electrons interact strongly with each other, there is almost no state to scatter to because all the states at the available energy are filled[1]. Therefore, only the correlation energies not included in the mean-field theory need to be considered. They are smaller and are easier to treat. However, a linear combination of core orbitals may have a large scattering energy comparable to the diagonal energy. This component can then be scattered above the Fermi surface and be relevant to the low energy physics. This effect is included in our method by the valence orbitals chosen at the initial RG steps.

A key strategy of RG technique is to include the perturbative effects of the discarded states[8, 22, 23]. When the perturbations are included recursively, the improvement may be substantial. The forced separation of energy scales will make the perturbation converges faster.

From the low energy spectrum, we can obtain many other properties of the model such as low temperature specific heat and susceptibility (if the model has spin degrees of freedom). Since we also know the wave functions of low lying states, correlation functions can also be obtained.

We now discuss several crucial techniques that make this calculation possible. (*i*) To obtain an expression for the expectation values, such as $\langle\alpha_1'|H|\alpha_1\rangle$, it is easiest to first write the hamiltonian in terms of the new orbitals f_k. (*ii*) In order to calculate certain residual interactions when going from one RG step to the next, we need to save the matrix elements of operators such as $\langle\alpha_l'|f_p^\dagger f_q|\alpha_l\rangle$ where $M - M_l < p, q \leq M - M_l$ and $|\alpha_l'\rangle$ and $|\alpha_l\rangle$ are the retained states. Only the sum $\sum_{pq} U_{pi} U_{qi} \langle\alpha_l'|f_p^\dagger f_q|\alpha_l\rangle$ is needed for the calculation. We save the computer memory by a factor of M when the sum is stored instead of individual matrix elements of $f_p^\dagger f_q$. (*iii*) Generalization to other fillings will be easy by allowing one of $f_{l,core}^\dagger$ and $f_{l,empty}^\dagger$ to mix with the kept states in some sequence commensurate with the filling.

4 Conclusions

In summary, we have formulated a RG procedure that is adaptive to the correlation energy. The procedure selects a sequence of orbitals. The interactions among the orbitals are argued to display a hierarchy of energy scales. The transformed hamiltonian is then treated by the numerical RG technique. Our method yields a systematic improvement over the HF theory and has a wide range of applicability. Our test calculation on an exactly solvable model shows that the new technique is substantially better than the HF theory, with typical error of a few percent for the low energy excitations. The method can be improved by including the perturbative effects of the discarded states.

Acknowledgements

I would like to thank Prof. Sriam Shastry for useful suggestions. This work is supported by the Office of Naval Research Contract No. N00014-95-1-0398 and by the National Science Foundation Grand No. DMR-9403201. The work is also support by the Center for Academic Computing at The Pennsylvania State University.

References

[1] V. F. Weisskoft, *Science* **113**, 101 (1951).

[2] *High Temperature Superconductivity*, edited by K. S. Bedell, *et al.* (Addison-Wesley, Redwood City, CA, 1990).

[3] *The Quantum Hall effect*, edited by Richard E. Prange and Steven M. Girvin, 2nd ed. (New York, Springer-Verlag, 1990).

[4] N. Grewe and F. Steglich in *Handbook on the Physics and Chemistry of Rare Earths* **14** (1991).

[5] R. Blankenbecler, D. J. Scalapino, and R. L. Sugar, *Phys. Rev. D* **24**, 2278 (1981).

[6] J. E. Hirsch, *Phys. Rev. B* **31**, 4403 (1985).

[7] Recently, there has been a renewed interest in application of RG techniques to correlated systems. See, for example, R. Shankar, *Rev. Mod. Phys.* **66**, 129 (1994).

[8] K. G. Wilson, *Rev. Mod. Phys.* **47**, 773 (1975).

[9] K. G. Wilson, in *Nobel Symposia-Medicine and Natural Sciences* (Academic Press, New York) **24**, 68 (1974); *Adv. Math.* **16**, 170 (1975).

[10] J.E. Hirsch, *Phys. Rev. B* **22** (1980) 5259.

[11] S. R. White, *Phys. Rev. Lett.* **69** (1992) 2863, *Phys. Rev. B* **48** (1993) 10345.

[12] S. Liang and H. Pang, *Phys. Rev. B* **49** (1994) 9214.

[13] Shoudan Liang, *J. Phys. Chem. Solids*, **56** 1729 (1995).

[14] Shoudan Liang, *Phys. Rev. Lett.*, **75**, 3493 (1995).

[15] The off-diagonal terms in m_i are exactly zero when the density of the **k**-states within each shell as a function of the angular variable θ_j is constant. On a finite lattice, the density fluctuate around a constant so the off-diagonal terms will not be exactly zero, but will be very small.

[16] C. N. Yang and C. P. Yang, *Phys. Rev.* **150** 321 (1966).

[17] G. H. Golub and C. F. Van Loan, *Matrix Computations* (The Johns Hopkins University Press, Baltimore) 1989.

[18] S. Liang, to be published.

[19] A. C. Wahl and G. Das, in *Methods of Electronic Structure Theory*, H. F. Schaefer III ed, (Plenum Press, New York), 1977.

[20] P. W. Anderson, *Science* **235** (1987) 1196.

[21] For a review on Hubbard model, see E. Dagotto, *Rev. Mod. Phys.* **66**, 763 (1994).

[22] T. Tokuyasu, M. Kamal, G. Murthy, *Phys. Rev. Lett.* **71**, 4202 (1993).

[23] M. Lepetit and E. Manousakis, *Phys. Rev. B* **48**, 1028 (1993).

Optimized Trial Functions Applied to Static and Dynamic Equilibrium Critical Phenomena, and Quantum Mechanical van der Waals Clusters

M.P. Nightingale

Department of Physics, University of Rhode Island, Kingston, RI 02881, USA

Monte Carlo computation of eigenvalues is discussed for quantum mechanical Hamiltonians, and transfer and stochastic matrices. In all cases, computations exploit optimized trial states, which results in a sizable reduction of the statistical variance of these Monte Carlo methods.

I. INTRODUCTION

Many computational problems can be reduced to the computation of eigenvalues of operators. Examples of operators discussed in this paper are quantum mechanical Hamiltonians, transfer matrices, and stochastic (Markov) matrices. Numerical computation of the eigenvalue spectrum of the full operator, or its approximate counterpart in a truncated representation of the state space, has been employed traditionally for this purpose, but more recently Monte Carlo eigenvalue methods have become practical alternatives.

A particularly appealing feature of Monte Carlo eigenvalue methods is that the stochastic process can be used to estimate *only corrections* to already sophisticated approximations. More specifically, these methods can de designed so that statistical errors vanish in the ideal case that the optimized trial states, which are pivotal in this approach, are exact eigenstates. In practice, one obviously cannot realize this ideal, but one can get close since there is much flexibility in the choice of trial functions, a flexibility which can be exploited to incorporate important physics in the approximation from the outset.

In applications, much work goes into the the design and optimization of these trial states and this paper reviews various examples, *viz.* the computation of the ground state energy of van der Waals clusters, and the computation of critical exponents of static and dynamic equilibrium models in two and three dimensions. The examples discussed here are selected from work performed by the author with various collaborators[1–9] and this paper is not intended as a comprehensive review of the field.

II. TRIAL FUNCTION OPTIMIZATION

First we consider the general principle[10,11] of Monte Carlo optimization of a trial vector for the dominant eigenstate of an operator $\mathbf{T}$. Here, the term "dominant" is used loosely and its meaning depends on the application. As

Springer Proceedings in Physics, Volume 82
Computer Simulation Studies in Condensed-Matter Physics IX
Eds.: D. P. Landau, K.K. Mon, H. -B. Schüttler

mentioned, three types of applications will be discussed in this paper: quantum mechanics, and static and dynamic equilibrium statistical mechanics. If the operator $\mathbf{T}$ is a quantum mechanical Hamiltonian $\mathcal{H}$, the "dominant" state is the ground state, which strictly speaking is only the true dominant state of, e.g., the imaginary-time evolution operator $\exp t\mathcal{H}$. In the statics case, $\mathbf{T}$ is the transfer matrix, and the dominant eigenstate truly is the state with the eigenvalue of largest magnitude. In the application to stochastic processes satisfying detailed balance, the largest eigenvalue is unity and the associated eigenstate is also known: the Boltzmann distribution. In this case, "dominant state" denotes the dominant non-trivial state, i.e., the state with the second largest eigenvalue.

For simplicity of presentation we assume that the operator $\mathbf{T}$ is hermitian. In practice, the method is applied to non-hermitian operators as well. That is, transfer matrices are used for systems with helical boundary conditions. Owing to the helicity, the transfer matrices have the property

$$\mathbf{T}^{\dagger} = \mathbf{RTR}, \tag{1}$$

where $\mathbf{R}$ is a reflection operator, i.e., $\mathbf{R}^2 = \mathbb{1}$. This property implies that a simple relation exists between left and right eigenstates, which is usually not the case for non-hermitian operators. As a consequence, the discussion below can be immediately generalized to tranfer matrices, but to simplify presentation we omit the technical details.

Consider a trial state $|\psi_{\mathrm{T}}\rangle$ depending on a set of optimization parameters. Optimal values for the parameters can, in principle, be found by minimization of the quantity

$$\chi^2 = \frac{\langle\psi_{\mathrm{T}}|(\mathbf{T} - \langle\mathbf{T}\rangle)^2|\psi_{\mathrm{T}}\rangle}{\langle\psi_{\mathrm{T}}|\psi_{\mathrm{T}}\rangle}, \tag{2}$$

where

$$\langle\mathbf{T}\rangle = \frac{\langle\psi_{\mathrm{T}}|\mathbf{T}|\psi_{\mathrm{T}}\rangle}{\langle\psi_{\mathrm{T}}|\psi_{\mathrm{T}}\rangle}. \tag{3}$$

We note that χ^2 assumes its minimum value, zero, for any eigenstate of the operator $\mathbf{T}$, not just the dominant state.

In practice, the variance χ^2 cannot be evaluated directly and a Monte Carlo estimate is used instead. This can be done by generating a sample $\mathcal{S}$ of configurations s drawn from a suitable distribution with relative weights $\langle s|\psi_{\mathrm{T}}{}^{(0)}\rangle^2$, as given, for example, by a trial state defined by initial guesses of the optimization parameters. Given this sample $\mathcal{S}$, one can evaluate

$$\chi^2 \approx \frac{\sum_{s\epsilon\mathcal{S}}[t(s) - \bar{t}]^2 w_s^2}{\sum_{s\epsilon\mathcal{S}} w_s^2}. \tag{4}$$

Here the w_s are weights defined by $w_s = \langle s|\psi_{\mathrm{T}}\rangle/\langle s|\psi_{\mathrm{T}}{}^{(0)}\rangle$ and $\bar{t}$ is the corresponding weighted sample average of the *configurational eigenvalue,* which is called the *local energy* in real-space quantum Monte Carlo applications,

$$t(s) = \frac{\langle s|\mathbf{T}|\psi_{\mathrm{T}}\rangle}{\langle s|\psi_{\mathrm{T}}\rangle}. \tag{5}$$

The configurational eigenvalue $t(s)$ should be calculable without explicit summation or integration over all states of a complete set inserted between $\mathbf{T}$ and $|\psi_{\mathrm{T}}\rangle$, a condition which is satisfied *e.g.* if $\mathbf{T}$ is sparse or quasi-local. In this case, optimal parameters can be found efficiently with a modified version[12] of the Levenberg-Marquardt algorithm, designed to minimize a sum of squares, such as the expression given in Eq. (4). The crux of the method is that this procedure is applied to a fixed, small sample, which works efficiently as long as the sample is a good representation of $|\psi_{\mathrm{T}}\rangle$, the trial state defined by the current parameter estimates. How well the sample represents this state can be measured by the overlap

$$c = \frac{\langle \psi_{\mathrm{T}}|\psi_{\mathrm{T}}{}^{(0)}\rangle}{\sqrt{(\langle \psi_{\mathrm{T}}{}^{(0)}|\psi_{\mathrm{T}}{}^{(0)}\rangle\langle \psi_{\mathrm{T}}|\psi_{\mathrm{T}}\rangle)}}, \tag{6}$$

or in more practical terms

$$c^2 \approx \frac{(\sum_{s\epsilon\mathcal{S}} w_s)^2}{|\mathcal{S}|\sum_{s\epsilon\mathcal{S}} w_s^2}. \tag{7}$$

Sometimes, the weights w_s can be negative, in which case, w_s is replaced by $|w_s|$ and Eq. (6) should be modified accordingly. Whenever the overlap becomes too small, a new sample has to be generated, e.g., from the best current trial state, and the process is iterated until it converges.

III. BOSONIC VAN DER WAALS CLUSTERS

As a first example, we consider van der Waals clusters of N bosonic Lennard-Jones atoms. In reduced units, the Lennard-Jones pair potential has the form $v(r) = r^{-12} - 2r^{-6}$, and the only independent parameter in the Schrödinger equation is the reduced inverse mass m^{-1}, a quantity proportional to the square of the de Boer parameter, $h/\sigma\sqrt{m\epsilon}$, where σ and ϵ are the usual parameters of the Lennard-Jones potential and h is Planck's constant.

A trial wave function for the ground state of a bosonic cluster should be invariant under translation, rotation and particle permutation. The first two of these requirements can be satisfied by choosing as coordinates the interatomic distances r_{ij}, $1 \le i < j \le N$. The third condition, particle exchange symmetry, then can be imposed explicitly by considering only functions that are invariant under the $N!$ permutations of the indices of the r_{ij}.

In addition to these symmetry restrictions, the trial function has to be positive in the ground state. In principle, n-body correlations with all $n \le N$, should be incorporated in the trial wave function, but these many-body effects are expected to become progressively less important as n increases. Positivity and many-body correlations suggest that the trial function be written as

$$\log \psi_T = \sum_{(i,j)} u^{(2)}(r_{ij}) + \sum_{(i,j,k)} u^{(3)}(r_{ij},\ r_{jk},\ r_{ki}) + \cdots, \tag{8}$$

where the $u^{(n)}$ are real-valued, n-body functions.

In the design of trial wave functions, described in detail in Ref. 5, we followed the procedure developed by Umrigar *et al.*[10,11]. That is, the trial functions satisfy boundary conditions associated with (a) the collision of two atoms and (b) having one atom go off to infinity. The behavior of the wave function for the most likely configurations, which involve intermediate distances and require most of the variational freedom of the trial wave functions, is described by expanding the $u^{(n)}$ in polynomials of variables $\hat{r}_{ij} = f(r_{ij})$, where $f(r)$ is an optimizable function that approaches a constant for $r \to \infty$. Note that $u^{(2)}$ is exceptional in that in addtition to these polynomial terms, it has singular terms dictated by the boundary conditions.

Wave functions of this type can be used to study the importance of many-body effects for the quality of the trial wave function. The exact ground state energy E_0 and its variational estimate $\tilde{E}_0$ satisfy the inequality

$$\tilde{E}_0 - \chi \leq E_0 \leq \tilde{E}_0. \tag{9}$$

This suggests that the quality of the trial wave function ψ_T be measured by

$$Q = -\log_{10} \frac{\chi}{|E_0|}. \tag{10}$$

The limit $Q \to \infty$ corresponds to an *exact* solution of the time-independent Schrödinger equation. In fact, there exists a tighter bound, linear in χ^2 rather than χ, and we refer to Refs. 5 and 7 for details.

Our first numerical example addresses how the quality of the trial wave function improves as n body terms with progressively larger values of n are included. Figures 1 through 3 display results for Ar clusters of sizes three, four, and five. Q, a conservative estimate of the quality of the wave functions as defined in Eq. (10), is plotted versus the power P of the polynomials used in the expansion of Eq. (8) for different values of n. The bottom curve, starting at $P = 1$, is for the case in which only two-body interactions are present ($n = 2$). As shown, the quality levels off at a fixed value of Q with increasing P, an indication that the absence of three-body terms is the dominant source of variance of the local energy. In the next curve segment, three-body terms are added ($n = 3$). Here P is redefined to denote the order of three-body polynomial, which starts at $P = 2$; in the second curve segment the order of the polynomial that describes two-body effects is kept constant at the highest P-value used in the previous segment of the curve. This process is repeated for n-body terms with increasing n until finally the complete N-body polynomial for the N-atom cluster is included in the trial function. At this point, the quality starts to go up roughly linearly with the order of

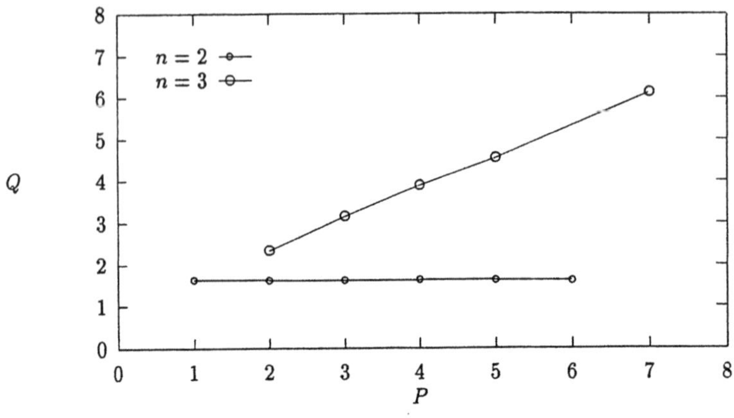

FIG. 1. Quality Q, a measure of the accuracy of the optimized trial function as defined in Eq. (10), as a function of the power P of two- and three-body polynomials, labeled by n, for Ar_3.

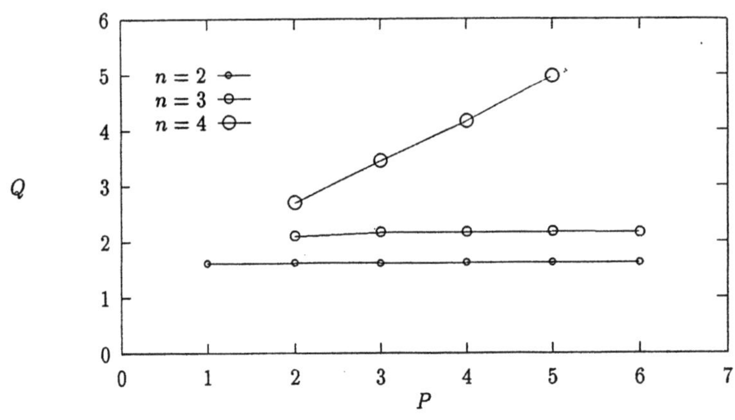

FIG. 2. Quality Q, a measure of the accuracy of the optimized trial function as defined in Eq. (10), as a function of the power P of two- , three- and four-body polynomials, labeled by n, for Ar_4.

the polynomial. Similar results were obtained for systems made of lighter atoms.[5] Here we only note that the order P of the four-body polynomials is not sufficiently high in any of the figures for the quality to have leveled off, as ultimately it must for five-atom clusters.

Figures 4, 5, and 6 are plots of the quality Q *vs.* the power P for each type and size of cluster for optimized wave functions including the full many-body polynomials. An interesting feature of these plots is that the Q *vs.*

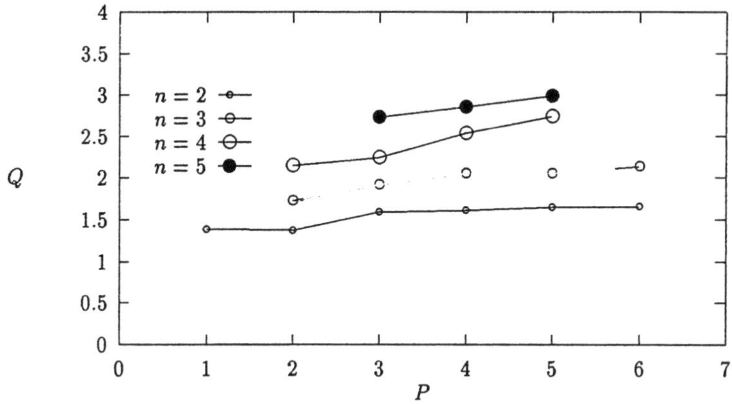

FIG. 3. Quality Q, a measure of the accuracy of the optimized trial function as defined in Eq. (10), as a function of the power P of two-, three-, four- and five-body polynomials, labeled by n, for Ar_5.

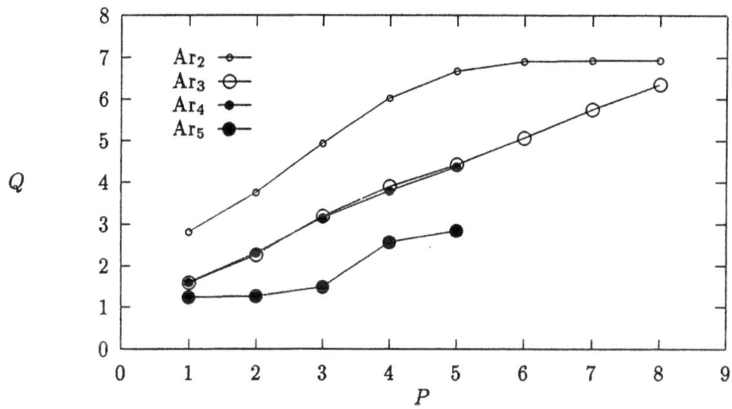

FIG. 4. Quality Q, a measure of the accuracy of the optimized trial function as defined in Eq. (10), as a function of power P of the complete N-body polynomial for argon clusters of sizes two through five. The curve for Ar_2 levels of because of noisy behavior of the numerical derivatives used in the computations. Apparent erratic behavior of the Ar_5 curve is presumably due to occasional suboptimal parameter choices for the trial wave function.

P curves for three- and four-atom ($N = 3, 4$) clusters almost coincide but that they are distinct from the curves for $N = 2$ and $N = 5$. The $N = 2$ clusters are unique in that the short-distance divergences in the local energy have been fully removed, while also the large-distance asymptotic properties of the trial function is superior in this case. Apart from this, the effect probably is geometric in nature: clusters of sizes $N \leq 4$ are fully symmetric

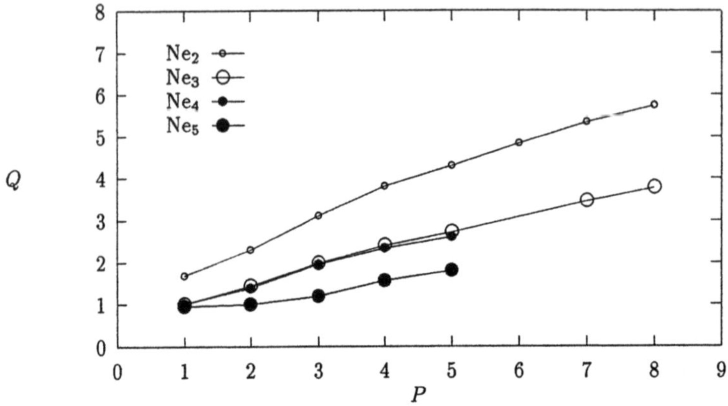

FIG. 5. Quality Q, a measure of the accuracy of the optimized trial function as defined in Eq. (10), as a function of power P of the complete N-body polynomial for neon clusters of sizes two through five.

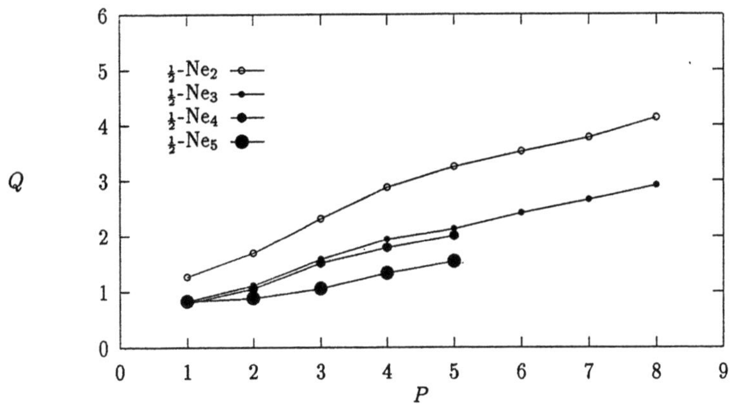

FIG. 6. Quality Q, a measure of the accuracy of the optimized trial function as defined in Eq. (10), as a function of power P of the complete N-body polynomial for $\frac{1}{2}$-neon clusters of sizes two through five.

in the classical configuration of minimum energy and can be characterized by a single inter-atomic distance. For $N = 5$ the cluster is frustrated: the classical ground state is a trigonal bipyramid (the faces of which are six congruent, isosceles triangles) and has three different inter-atomic distances.

In the process of employing the trial function optimization described in Section II, it is important to know what regions of configuration space contribute most to the variance χ^2, as defined in Eq. (2). For instance, it is useful to know if the quality of the wave function is limited by poorly satisfied boundary conditions or if the quality can be improved by adding more

FIG. 7. Density plot of the "local error" in the geometry described in the text. The two dots in the lower right-hand corner are the two in-plane vertices of the tetrahedron; the one in the upper left corner is the projection of the two out-of-plane vertices. The length of the tetrahedron edges is 1.3 and $m^{-1} = 0.16$. The darker the region, the more it contributes to χ^2. Note that the dark region in the lower right is a cut through the banana-shaped dark region in the upper left. The regions of the largest local error are the two symmetrically located regions of collision of four atoms. White lines are are cuts through nodal surfaces of the local error and have no physical significance.

variational parameters. Another possibility is that the wave function has too much variational freedom relative to the size of the sample over which it is optimized. This might show up in the form of unphysical peaks in the wave function, which might remain invisible in the variance of the local energy estimated from the relatively small samples used to optimize the trial function.

To help answer such questions, one can make density plots of the local error, the deviation of the local energy from its average and, as an illustration, we

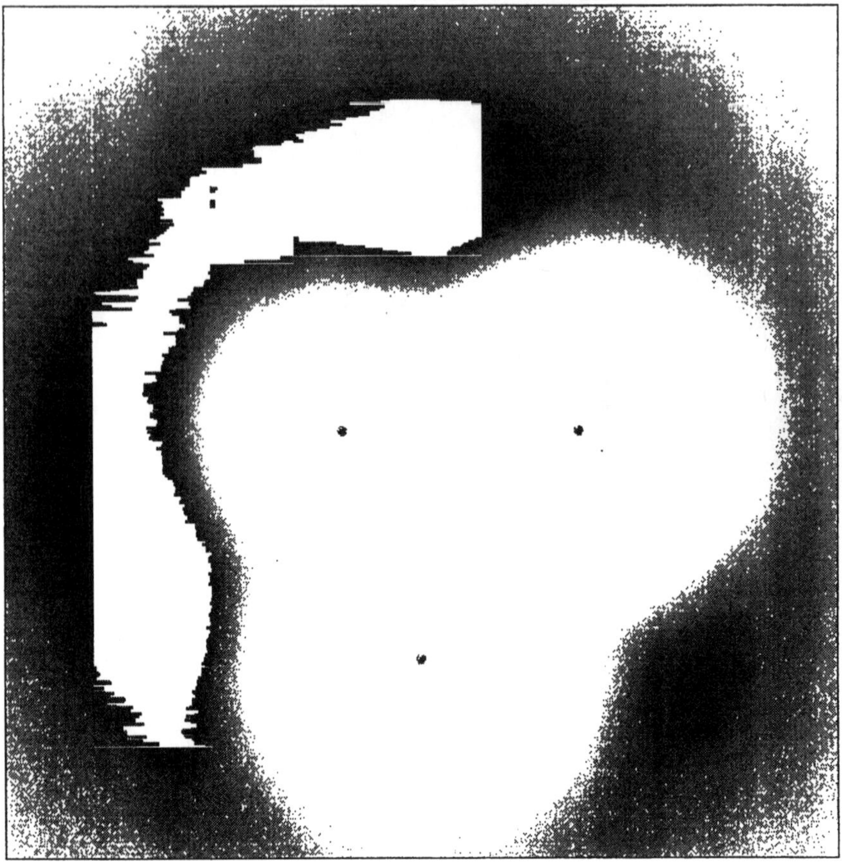

FIG. 8. Density plot of $|\psi_T|^2$ as a function of the position of the fifth particle in the geometry described in the text and the figure caption of Fig. 7. Dark regions correspond to high probability density. Note that the regions with the largest local error are contained in the region where the wave function becomes small because of repulsive potential at short range.

discuss a five-atom cluster. In fact, it is useful to superimpose color density plots of both the wave function and the local error, which contain more information than can be reproduced by the grey-scale plots reproduced in this paper. See Ref. 13 for the color graphics.

Obviously, the fact that the ground state wave function depends on $3N-6$ independent coordinate variables, seriously limits any graphical approach. For the five-atom clusters the following planar cut through configuration space is informative: four atoms are fixed at the vertices of a regular tetrahedron, while the fifth particle is located in a plane that contains two of these vertices and bisects the edge connecting the two remaining atoms.

Figures 7 and 8 strongly suggest which regions of configuration space contribute most to χ^2 and for the interpretation of the density plots the following convention should be used: zero intensity (white) corresponds to a minimum, while full intensity (black) corresponds to a maximum of the plotted function. Figure 7 represents the density plot of the weighted *local error*, defined as $|(\mathcal{E} - E_{\rm T})\psi_{\rm T}|$, as a function of the position of the fifth, wandering atom. [Note that the *local energy* $\mathcal{E}$ is defined by Eq. (5) with **T** replaced by the Hamiltonian.] Figure 8 shows the dependence of $\psi_{\rm T}^2$, on the position of the fifth atom while the other four atoms are fixed in the geometry mentioned above.

The conclusion drawn from inspection of the density plots is that the trial function fails particularly in regions where more than two atoms approach each other, and we see that of these the four-body collisions are worst at least in a local sense. Unfortunately, sofar we have not been able to produce trial wave functions without this flaw.

IV. STATICS AT THE CRITICAL POINT

In the next example, we discuss the use of optimized trial functions to compute the dominant eigenvalue of a transfer matrix. Note that the eigenvalue is of immediate physical significance, since its logarithm is proportional to the free energy.

We give a brief sketch of how the functional form of trial functions can be guessed for lattice systems, and we use as an illustration the two-dimensional Ising model. Again, the reader is referred for details to the original paper.[9]

Consider a simple-quadratic lattice of M sites, wrapped on a cylinder with a circumference of L lattice units. If helical boundary conditions are used, the one-spin transfer matrix is

$$T_{S,R} = e^{K(s_1 r_1 + s_1 r_L)} \prod_{i=1}^{L-1} \delta_{s_i, r_{i+1}}, \tag{11}$$

with $S = (s_1, s_2, \ldots, s_L)$ and $R = (r_1, r_2, \ldots, r_L)$, where the $s_i = \pm 1$ and $r_i \pm 1$. The conditional partition function of the lattice of M sites, subject to the condition that the spins on the left-hand edge are in state R, as illustrated in Fig. 9, is denoted $Z_M(R)$. One has

$$Z_{M+1}(S) = \sum_R T_{S,R} Z_M(R). \tag{12}$$

For $M \to \infty$ the vector of conditional partition sums $Z_M(R)$ form the dominant right eigenvector of the transfer matrix. An element of this eigenvector is represented by the graph on the right in Fig. 9, at least if the lattice segments are viewed as semi inifinite. Full circles indicate spins that have been summed over in the partition sum; the *fixed* surface spins are represented by the open circles. Each bond represents a factor $\exp(K s_i s_j)$. The left eigenvector, which is the one that has to be approximated by an optimized

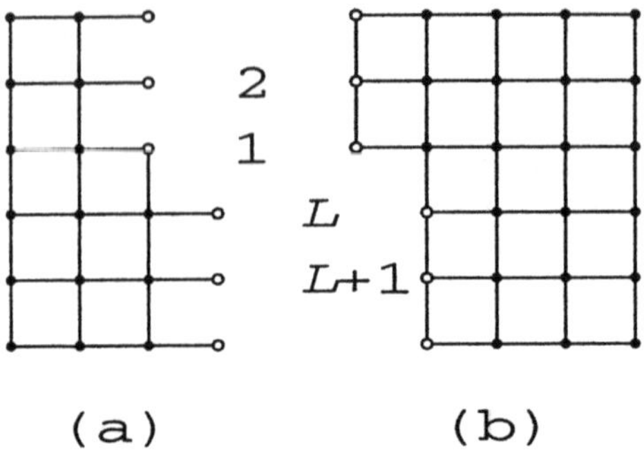

FIG. 9. Illustration of left and right eigenvectors of the transfer matrix. Open circles are the sites of fixed spins; closed circles have been summed over in the conditional partition function. The lattice segments are to be considered semi infinite.

trial vector, is represented by the graph on the left. In constructing a trial state, an effective Hamiltonian is required, to describe the interactions of the surface spins (open circles) in equilibrium with the bulk (full circles); the eigenvector is the corresponding vector of Boltzmann weights. The following is an obvious choice

$$\psi_{\mathrm{T}} = \exp \sum_{i,j}^{*} K_{ij} s_i s_j . \tag{13}$$

Here the asterisk indicates that the sum over all pair interactions has to be truncated for reasons of numerical efficiency, but of course, even with all pair interactions included, Eq. (13) is only an approximation. The coupling constants are the variational parameters. Note that the K_{ij} lack translational invariance, because of the step on the surface, apparent from Fig. 9. This causes technical problems requiring a solution outside the scope of this review.

As an illustration of the efficiency and flexibility of these trial vectors, we discuss the XY-Ising model. It consists of coupled Ising and planar rotator degrees of freedom on a simple-quadratic lattice. On each lattice site there are two variables: $s_i \pm 1$ and $\mathbf{n}_i$, a two-component unit vector. The Hamiltonian of this model is given by

$$H = -k_{\mathrm{B}} T \sum_{(i,j)} (A\, \mathbf{n}_i \cdot \mathbf{n}_j + B\, \mathbf{n}_i \cdot \mathbf{n}_j s_i s_j + C s_i s_j) , \tag{14}$$

where the summation is over nearest-neighbor pairs.

To illustrate the performance of transfer-matrix Monte Carlo algorithm we consider the special case $A = B$. For a discussion of the physics of this model the reader is referred to Ref. 6. The trial vectors discussed above for the Ising model have an immediate generalization

$$\psi_{\mathrm{T}} = \exp \sum_{i,j}^{*} (A_{ij}\, \mathbf{n}_i \cdot \mathbf{n}_j + B_{ij}\, \mathbf{n}_i \cdot \mathbf{n}_j s_i s_j + C_{ij} s_i s_j) . \tag{15}$$

TABLE I. Estimated eigenvalue and standard errors for the XY-Ising model. These data apply to the point ($A = 1.005, C = -0.2285$) [*cf.* Eq. (14)] on the line where Ising and XY transitions coincide[6]. Results are shown for various values of d_m, the maximum path length of the cutoff in Eq. (15). The results[9] are for a strip of width $L = 20$.

λ_0	σ	d_{max}	time (arbitrary units)
34.17406	0.0071	0	15
34.20875	0.0052	2	15
34.21658	0.0015	3	17
34.21418	0.00083	4	19
34.21384	0.00052	5	21
34.21366	0.00049	6	23
34.21379	0.00041	7	26

Table I shows estimates of the dominant eigenvalue of XY-Ising model for trial vectors truncated at various maximum values of d_{ij}, which essentially is the distance of the pairs (i, j) appearing in the sum over pair interactions. As can be seen by comparing the first and last lines of the table, the variance in the estimate of the eigenvalue is reduced by a factor 300 for a fixed number of Monte Carlo steps. Taking into account that the computer time per step doubles, this constitutes a speed-up by a factor of 150.

The truncation scheme introduced above for the Ising model is purely geometrical, and therefore carried over with only the obvious modifications to the XY-Ising model. The same is true for the three-dimensional models discussed below. It should, however, be noted that there are models and choices of transfer matrices to which this scheme does not apply. Ref. 4 contains a discussion and an example of such a case.

As a further illustration of the trial function optimization technique, we discuss applications to three-dimensional, simple-cubic $O(n)$ models for $n = 1$, 2 and 3, *i.e.*, the Ising, planar and Heisenberg models.

As mentioned above, transfer-matrix Monte Carlo is designed to compute the dominant eigenvalue λ_0 of the transfer matrix. The reduced free energy per site is $f = -\ln \lambda_0$ and one can calculate the interface free energy as the difference in free energy of two systems: one with ferromagnetic and the other with antiferromagnetic interactions, if the dimensions are chosen so as to force an interface in the antiferromagnetic system. For $L \times L \times \infty$ systems with helical boundary conditions, this means that L has to be even.

Renormalization group theory predicts that the values of Δ, the reduced interface free energy per lattice site, as a function of coupling K and system sizes L collapse onto a single curve, at least close to the critical point K_c and for sufficiently big systems. In terms of the non-linear thermal scaling field

$$u(K) = K - K_c + a(K - K_c)^2 + \dots, \tag{16}$$

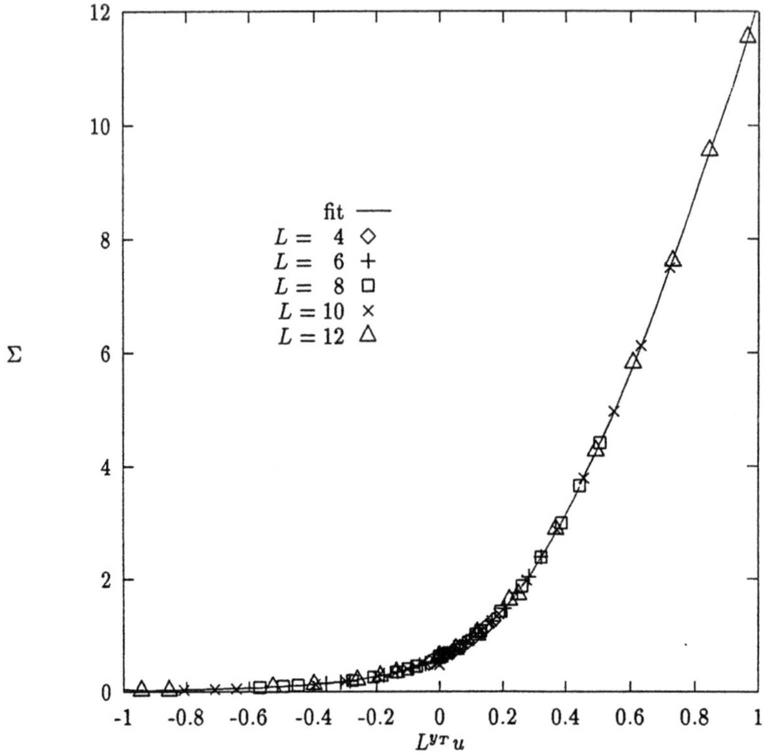

FIG. 10. Finite-size scaling plot for the interface free energy of the three dimensional Ising model.

this curve $\Sigma(x)$ is of the form

$$\Delta(u, L) = L^{1-d}\Sigma(L^{y_T}u), \tag{17}$$

for a d-dimensional system with a thermal scaling exponent y_T. The function Σ can be expanded in a series:

$$\Sigma(x) = \sum_{l=0}^{\infty} \sigma_l x^l. \tag{18}$$

The scaling plot is obtained by treating K_c, y_T, and the σ_l as fitting parameters and the result for the three-dimensional Ising model is shown in Fig. 10. To check if the system sizes are in the asymptotic, finite-size scaling regime given the statistical accuracy of the Monte Carlo data, fits can be made both with and without the $6 \times 6 \times \infty$ data. The results are: $K_c = 0.22162 \pm 0.00002$ and $y_T = 1.584 \pm 0.004$ using data with $L = 6$ through 12; and $K_c = 0.22167 \pm 0.00004$ and $y_T = 1.584 \pm 0.014$ if the $L = 6$ are omitted. These results agree well with those of other methods (see e.g.

Ref. 14–16 and references therein) which are in the vicinity of $K_c = 0.221655$ (with a margin of error of about 10^{-6}) and $y_T = 1.586$ (with a precision of a few times 10^{-3}).

Analysis of the interface free energy of $L \times L \times \infty$ systems, as illustrated in Fig. 10, suggests that corrections to scaling are remarkably small compared to the corrections that haunt standard Monte Carlo analyses[17] for $L \times L \times L$ systems.

A similar analysis can be performed for the three-dimensional planar model and yields the critical coupling $K_c = 0.45410 \pm 0.00003$ for system sizes $L = 6$ to 12, and $K_c = 0.45413 \pm 0.00005$ for $L = 8$ to 12. These values are close to results from series expansions[18,19] $K_c = 0.45386$ and standard Monte Carlo calculations[20] $K_c = 0.4531$ (no error estimate quoted). Also the estimates for the temperature exponent, *viz.* $y_T = 1.491 \pm 0.003$ for $L \geq 6$ and $y_T = 1.487 \pm 0.006$ for $L \geq 8$ agree well with other results $y_T = 1.495 \pm 0.005$.[21]

The calculations for the Heisenberg case were concentrated in a narrow range of temperatures around the critical point, and no attempt was made to determine the thermal exponent accurately and independently. If, in making the scaling plot, the latter is used as a fitting parameter anyway, the result equals, well within the statistical accuracy, the coupling-constant-expansion estimate[21] $y_T = 1.418$. Fixing the y_T to this value, the latter yields the following results: $K_c = 0.69291 \pm 0.00004$ for system sizes $L = 6$ to 12, and $K_c = 0.69294 \pm 0.00008$ for $L = 8$ to 12. These values are close to results from series expansions[22]: $K_c = 0.6916$, and more recently[19]: $K_c = 0.69294$; and from Monte Carlo calculations[23]: $K_c = 0.693035 \pm 0.000037$. The difference with the transfer matrix Monte Carlo result with the $L = 6$ data included might be a small finite-size effect.

Finally, it should be noted that, although in each of these three-dimensional models the finite-size effect appears to be small for $L \geq 6$, it is substantial for $L = 4$. For this reason the $L = 4$ data were not included in the fits.

V. DYNAMICS AT THE CRITICAL POINT

A problem with a long history of intractabilty is the computation of the dynamic critical exponent z to an accuracy sufficient to address meaningfully questions of universality. This is true even for systems as simple as the two-dimensional Ising model, although recently considerable progress appears to have been made.

Optimized trial vectors have made it possible to perform finite-size studies quite analogous to the ones discussed in the previous section. For the dynamic problem the analogue of the tranfer matrix is the stochastic (Markov) matrix **P** governing the dynamics; the interface free energy has as its analogue the inverse auto-correlation time. The only difference with the static case is that the dominant eigenvalue and eigenvector are known: the eigenvalue is unity and the corresponding eigenvector is the Boltzmann distribution, both by construction of the stochastic matrix

The correlation time τ_1 (in units of single transitions as described by the stochastic matrix $\mathbf{P}$) is determined by the second-largest eigenvalue λ_L of $\mathbf{P}$:

$$\tau_1^{-1} = -\ln \lambda_L . \tag{19}$$

For a system symmetric under spin inversion, the corresponding eigenvector is expected to be antisymmetric.

The element $P(s'|s)$ of the stochastic matrix denotes the probability of a single-spin flip transition from configuration s to s'. The matrix $\mathbf{P}$ satisfies detailed balance, and consequently, denoting by $\psi_{\mathrm{B}}(s)^2$ the Boltzmann weight of configuration s,

$$\hat{P}(s'|s) = \frac{1}{\psi_{\mathrm{B}}(s')} P(s'|s) \psi_{\mathrm{B}}(s) \tag{20}$$

defines a symmetric matrix $\hat{\mathbf{P}}$. The results discussed below were obtained for the heat bath or Yang[24] transition probabilities with random site selection.

For an arbitrary trial state $|\psi_{\mathrm{T}}\rangle$ and time t an effective eigenvalue $\lambda_L^{(t)}$ can be defined by

$$\lambda_L^{(t)} = \frac{\langle \hat{\mathbf{P}}^{t+1} \rangle_{\psi_{\mathrm{T}}}}{\langle \hat{\mathbf{P}}^{t} \rangle_{\psi_{\mathrm{T}}}}, \tag{21}$$

where $\langle \cdot \rangle_{\psi_{\mathrm{T}}}$ denotes the expectation value in the state $|\psi_{\mathrm{T}}\rangle$. In the generic case, the effective eigenvalue converges for $t \to \infty$ to the dominant eigenvalue with the same symmetry as the trial state $|\psi_{\mathrm{T}}\rangle$; for finite projection time t, the effective eigenvalue has an exponentially vanishing, systematic error. It turns out that for any given trial state one can compute the right-hand side of Eq. (21) with standard Monte Carlo methods, since it can be written as the ratio of two correlation functions.

It turns out that straighforward generalization of the ideas presented in Section IV, *viz.* a short-distance expansion, fails to produce a high-quality optimized trial vector, even when higher-order spin interactions are included. Analysis of the exact eigenvectors $\mathbf{P}$ for $L \leq 5$ systems shows that the elements are given approximately by the Boltzmann factor times a function of the magnetization. Thus, one is led to trial functions defined in terms of the energy, and long-wavelength components of the Fourier transform of the spin configuration, the zero momentum component m_0 of which is just the magnetization per site. After much experimentation, the following form was selected

$$\psi_{\mathrm{T}}(s) = \tilde{\psi}_{\mathrm{B}}(s)\, \psi^{(+)}(s)\, \psi^{(-)}(s), \tag{22}$$

where $\psi^{(\pm)} \to \pm\psi^{(\pm)}$ under spin inversion, so that the trial function is antisymmetric under this transformation. The tilde in $\tilde{\psi}_{\mathrm{B}}$ indicates that the temperature is a variational parameter, but the optimal value of this variational temperature turns out to be virtually indistinguishable from the true temperature. The $\psi^{(\pm)}$ were chosen to be of the form

TABLE II. Estimated auto-correlation time τ_1 for $L \times L$ Ising systems.

L	τ_1	error
5	3435.28	0.07
6	7446.59	0.25
7	14279.3	0.63
8	25053.0	1.4
9	41089.2	2.9
10	63928.7	5.7
11	95288.4	9.1
12	137187	15
13	191717	22
14	261342	34
15	348602	61

$$\psi^{(+)} = \sum_{\mathbf{k}} a_{\mathbf{k}}(m_0^2) m_{\mathbf{k}}^{(+)} + m_0 \sum_{\mathbf{k}} b_{\mathbf{k}}(m_0^2) m_{\mathbf{k}}^{(-)} \tag{23}$$

$$\psi^{(-)} = m_0 \sum_{\mathbf{k}} c_{\mathbf{k}}(m_0^2) m_{\mathbf{k}}^{(+)} + \sum_{\mathbf{k}} d_{\mathbf{k}}(m_0^2) m_{\mathbf{k}}^{(-)}, \tag{24}$$

where the index $\mathbf{k}$ runs through a small set of multiplets of four or less long-wavelenght wave vectors defining the $m_{\mathbf{k}}^{(\pm)}$, translation and rotation symmetric sums of products of Fourier transforms of the local magnetization; the values of $\mathbf{k}$ are selected so that $m_{\mathbf{k}}^{(-)}$ is odd and $m_{\mathbf{k}}^{(+)}$ is even under spin inversion; the coefficients $a_{\mathbf{k}}$, $b_{\mathbf{k}}$, $c_{\mathbf{k}}$ and $d_{\mathbf{k}}$ are polynomials of second order or less in m_0^2. The degrees of these polynomials were chosen so that no terms occur of higher degree than four in the spin variables. It suffices to optimize approximately forty parameters for the trial functions used in this expample. Table II shows the values and standard errors of the single-spin-flip auto-correlation time computed with these trial functions.

To obtain an estimate for the critical exponent z, only a finite-size scaling analysis *at* the critical point is required. This can be done by fitting the data for the auto-correlation time to the form

$$\tau_1 = L^{z+2} \sum_{k=0}^{n_c} \alpha_k L^{2k}, \tag{25}$$

with z and the α_k as fitting parameters, and n_c as a cutoff which can be varied to check the convergence. There is no compelling theoretical justification for these particular correction terms other than that they apply to the static two-dimensional Ising model and provide a conservative extrapolation scheme. The result obtained from this analysis is $z = 2.1665 \pm 0.0012$ where the error quoted is the a two-sigma error as estimated from the least-squares fit. This appears to be the most precise estimate to date, and the reader is referred to Ref. 8 for a more detailed discussion of the analysis and an explicit comparisson with other work.

ACKNOWLEDGMENTS

This research was supported by the (US) National Science Foundation through Grant # DMR-9214669, by the Office of Naval Research and by the NATO through Grant # CRG 910152. This research was conducted in part using the resources of the Cornell Theory Center, which receives major funding from the National Science Foundation (NSF) and New York State, with additional support from the Advanced Research Projects Agency (ARPA), the National Center for Research Resources at the National Institutes of Health (NIH), IBM Corporation, and other members of the center's Corporate Research Institute.

References

[1] M. P. Nightingale and R. G. Caflisch, in "Monte Carlo calculation of eigenvalues of transfer matrices," in *Computer Simulation Studies in Condensed Matter Physics,* D. P. Landau and H. B. Sch"ttler (eds.),p. 208-213 (Springer, Berlin 1988).

[2] M. P. Nightingale, in *Proceedings of the Third International Conference on Supercomputing,* L. P. Kartashev and S. I. Kartashev (eds.) vol. I, 427-436 (1988).

[3] M.P. Nightingale and H.W.J. Blöte, Phys. Rev. Lett. **60**, 1662 (1988).

[4] E. Granato and M.P. Nightingale, Phys. Rev. B **48**, 7438 (1993).

[5] A. Mushinski and M.P. Nightingale, J. Chem. Phys. **101**, 8831 (1994).

[6] M.P. Nightingale, E. Granato and J.M. Kosterlitz, Phys. Rev. B **52**, 7402 (1995).

[7] M. Meierovich, A. Mushinski and M.P. Nightingale, J. Chem. Phys. in print, and URL http://xxx.lanl.gov /abs/chem-ph/9512001.

[8] M.P. Nightingale and H.W.J. Blöte, Phys. Rev. Lett. **76** 4548, 1996, also see URL http://xxx.lanl.gov/abs/cond-mat/9601059.

[9] M.P. Nightingale and H.W.J. Blöte, Phys. Rev. B **54** 1001, 1996, URL http://xxx.lanl.gov /abs/cond-mat/9602089.

[10] C.J. Umrigar, K.G. Wilson and J.W. Wilkins, Phys. Rev. Lett. 60, 1719, 1988.

[11] C.J. Umrigar, K.G. Wilson and J.W. Wilkins, in *Computer Simulation Studies in Condensed Matter Physics, Recent Developments,* edited by D.P. Landau K.K. Mon and H.B. Schüttler, Springer Proc. Phys. (Springer, Berlin, 1988).

[12] M.P. Nightingale and C.J. Umrigar, unpublished. For a test version of the program contact us at cyrus@tc.cornell.edu or nigh@repeter.phys.uri.edu.

[13] See URL http://www.phys.uri.edu/people/mark_meierovich/visual/Main.html for an informal presentation.

[14] H.W.J. Blöte, E. Luijten and J.R. Heringa, J. Phys. A **28**, 6289 (1995).
[15] H.W.J. Blöte, J.R. Heringa, A. Hoogland, E.W. Meyer and T.S. Smit, preprint (1996).
[16] R. Gupta and P. Tamayo, preprint; to appear in Int. J. Mod. Phys (1996).
[17] A.M. Ferrenberg and D.P. Landau, Phys. Rev. B **44**, 5081 (1991).
[18] M. Ferer, M.A. Moore and M. Wortis, Phys. Rev. B **8**, 5205 (1973).
[19] K. Ohno, Y. Okabe and A. Morita, Prog. Theor. Phys. **71**, 714 (1984).
[20] D.P. Landau, R. Pandey and K. Binder, Phys. Rev. B **39**, 12302 (1989).
[21] J.C. Le Guillou and J. Zinn-Justin, Phys. Rev. B **21** 3976 (1980)
[22] D.S Ritchie and M.E. Fisher, Phys. Rev. B 5, 2668, 1972.
[23] K. Chen, A.M. Ferrenberg and D.P. Landau, J. Appl. Phys. **73**, 5488 (1993).
[24] C.P. Yang, Proc. Symp. Appl. Math. **15**, 351 (1963).

Part III

Contributed Papers

Path-Integral Monte Carlo Simulation for H_2 and D_2 Adsorbed on Graphite

M. Kreer and P. Nielaba

Institut für Physik, KoMa 331, Universität Mainz,
D-55099 Mainz, Germany

Abstract. Molecular layers are very good realizations of two dimensional systems. Hydrogen molecules H_2, HD, D_2 adsorbed on graphite are excellent model systems for investigating the influence of substrate fields and of quantum effects on phase transitions. At a coverage of a complete commensurable layer in the $\sqrt{3} \times \sqrt{3}$ $R30^o$ structure experiments showed an anomalous effect, the system with the lighter H_2 molecules has a higher order–disorder transition temperature compared to the system with the heavier D_2 molecules. By a combination of path integral Monte Carlo and finite size scaling techniques we analyze this effect. In detail we study the order parameter and the cumulants and discuss the impact of the effective "size" of the quantum particles on the location of the critical temperature.

Two–dimensional (2D) layers at surfaces have become an interesting field of research during the last decade [1]. Most of the phase transitions in these systems occur at fairly low temperatures, and for many aspects of the behavior quantum effects need to be considered.

In an interesting study [2] Freimuth and Wiechert studied the phase diagrams of H_2 and D_2 molecules adsorbed on graphite. At low temperatures and coverages below the $\sqrt{3} \times \sqrt{3}$–monolayer coverage $\rho_{\sqrt{3}}$ they found phase coexistence of a gas phase with a $\sqrt{3} \times \sqrt{3}$- ordered phase. The coexistence region ends in a tricritical point at a temperature T_{tri}, with $T_{tri}(H_2) < T_{tri}(D_2)$. Above T_{tri} the phase transition from the ordered to the disordered phase is of second order and the transition temperature T_c increases with the coverage. An interesting effect was found at $\rho = \rho_{\sqrt{3}}$. Here $T_c(D_2)$ for the D_2 system is smaller than for the H_2–system, with $T_c(H_2) - T_c(D_2) \approx 2.5\ K$. Since this behavior is different from that at the tricritical temperatures, and since one would perhaps expect a lower transition temperature for the system with the lighter particles, this phenomenon is sometimes named an anomaly. In order to analyze this anomaly we performed PIMC simulations for both systems. We approximate all N molecules as spherical particles interacting with a Lennard–Jones potential, and since the energy difference from the rotational ground state to the first excited state is in the order of 100 K we assume that the particles only occupy the rotational ground state at the temperatures considered ($T < 20\ K$).

The Hamiltonian of this system is $H = T_{kin} + V_{LJ} + V_{surf}$, T_{kin} is the kinetic energy, V_{LJ} is the pair interaction ($\sigma = 2.96$ Å$, \varepsilon = 36.7\ K$ [3]), and

Springer Proceedings in Physics, Volume 82
Computer Simulation Studies in Condensed-Matter Physics IX
Eds.: D. P. Landau, K.K. Mon, H. -B. Schüttler

V_{surf} is the interaction potential [4] between an adsorbate particle at the position (x, y, z) and all other substrate particles, $V_{surf}(x, y, z) = E_0(z) + E_1(z) f_1(x, y)$. V_{surf} contains a laterally averaged contribution $E_0(z)$, which has only a z dependence (z is the distance from the surface) and it has a term $E_1(z) f_1(x, y)$ which depends on the position parallel to the surface, $E_1(z) = (2\pi\varepsilon_{HC}\sigma_{HC}^6/a_s)\{(\sigma_{HC}^6/30)(g_1/2z)^5 K_5(g_1 z) - 2(g_1/2z)^2 K_2(g_1 z)\}$,

$$f_1(x, y) = -2\{cos\frac{2\pi}{a}(x + \frac{y}{\sqrt{3}}) + cos\frac{2\pi}{a}(x - \frac{y}{\sqrt{3}}) + cos\frac{4\pi}{a}(\frac{y}{\sqrt{3}})\}, \quad (1)$$

with $g_1 = 4\pi/a\sqrt{3}$, $a_s = 5.24\ \mathring{A}^2$, $\varepsilon_{HC} = 32.05\ K$, $\sigma_{HC} = 3.18\ \mathring{A}$ [4]. K_n is the modified Bessel function of kind n and $a = 2.456\mathring{A}$ is the graphite lattice constant. The model is defined on a $2D$ triangular lattice (periodic boundary conditions) with side length S, the simulations are done at a coverage $\bar{\rho}$. The partition function is given by: $Z_N = Tr \exp -\beta \sum [p_j^2/2m_j + V_{surf}(r_j) + V_{LJ}(\{r_j\})]$, with the momentum p_j, position r_j and mass m_j of the jth particle. After using the Trotter product formula the PIMC partition function with Trotter dimension P (with $r_j^{(P+1)} = r_j^{(1)}$) is:

$$\begin{aligned} Z_N &= \lim_{P\to\infty} \left(\frac{mP}{2\pi\hbar^2\beta}\right)^{\frac{NP}{2}} \prod_{j=1}^{N} \prod_{l=1}^{P} \left\{ \left[\int_{-\infty}^{\infty} dr_j^{(l)} \right] \right\} \\ &\times \exp\left[-\beta \sum_{i<j}^{N} \sum_{l=1}^{P} \left\{ \frac{1}{P} V_{LJ}(r_j^{(l)}, r_i^{(l)}) \right\} \right] \\ &\times \exp\left[-\beta \sum_{j=1}^{N} \sum_{l=1}^{P} \left\{ \frac{mP}{2\hbar^2\beta^2} [r_j^{(l)} - r_j^{(l+1)}]^2 + \frac{1}{P} V_{surf}(r_j^{(l)}) \right\} \right] . \quad (2) \end{aligned}$$

Histograms P_L for the density and the order parameter ϕ_c of the commensurate structure were investigated in subsystems of the size $L \times L$. The phase transition temperature T_c was determined with finite size scaling methods in particular through determining the intersection of the Binder - cumulant [5] $U_L(\phi_c) = 1 - < \phi_c^4 >_L / 3 < \phi_c^2 >_L^2$, by variation of the temperature: $U_L(T_c) = U_{L'}(T_c)$ for $L \neq L'$. $< \phi_c^k >_L := \int \phi_c^k P_L(\phi_c) d\phi_c$ is the kth moment of the order parameter distribution in subsystems of the size $L \times L$. A typical run ($N = 108, P = 25$) with 10^5 Monte Carlo steps took about 10 CPU hours on a CRAY YMP.

At $\bar{\rho} = \rho_{\sqrt{3}}$ in the quantum treatment the ordered $\sqrt{3} \times \sqrt{3}$ phase is found in contrast to the classical case. In Fig. 1 the cumulants for different subsystem sizes are shown for H_2 and D_2 at a coverage $\bar{\rho} = \rho_{\sqrt{3}}$, clearly the transition temperature for D_2, which is found by the cumulant intersection point, is smaller compared to the H_2 system, the temperature difference between both critical temperatures is about 3 K, in rough agreement with the experimental findings. At a given temperature the average "extent" (diameter) of the quantum chains in case of H_2 molecules is bigger compared to the D_2 molecules, in agreement with the de Broglie wavelengths of the particles.

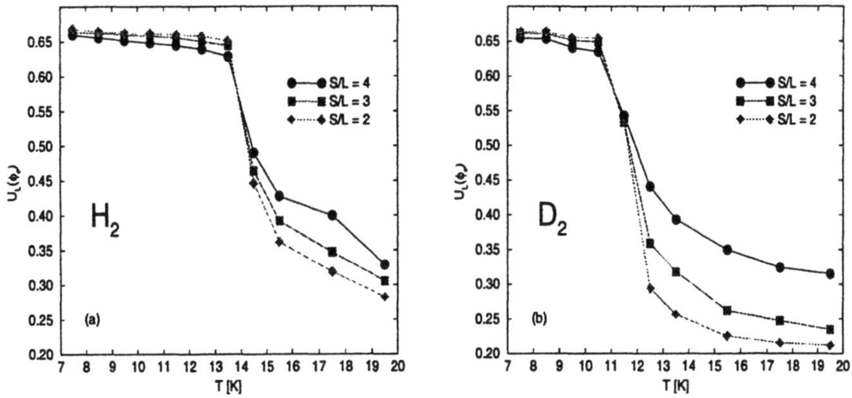

Figure 1: Cumulants of the order parameter ϕ_c for the $\sqrt{3} \times \sqrt{3}$ structure for H_2 (a) and D_2 (b) adsorbed on graphite ($S = 18\ a$, $N = 108$, $P = 25$). Points: PIMC results for different subsystem sizes L, lines are for visual help.

Thus the delocalization of H_2 molecules is more pronounced compared to D_2 molecules due to the larger de Broglie wavelength of H_2 molecules. As a result the effective repulsion of two H_2 particles is stronger and so the average distance between two H_2 particles is larger favoring the ordered structure. This may explain why the system with the lighter particles has a higher transition temperature to the disordered phase.

A question, which cannot be answered by experimental techniques, is how the behavior of the system changes, when the mass of the particles increases beyond the value of the D_2 particles, approaching the classical limit. In the PIMC we can study the system properties by modifying the mass and keeping the other interaction parameters fixed. In Figure 2 we see the results for the subsystem density distributions at the average density of (2/3) of the monolayer density ($\bar{\rho} = (2/3)\rho_{\sqrt{3}}$). For adsorbed particles with a mass $M = 1.5\ m_{H_2}$ we find at high temperatures ($T > 11.5K$) a density distribution peaked around the average density. At low temperatures ($T < 11.5K$) we obtain a doubly peaked density distribution with a peak at the low density gas phase and one peak at the monolayer density ($\rho = \rho_{\sqrt{3}}$). This shows that for masses in the range of the H_2 and D_2 molecules at low temperatures there is a phase coexistence between the gas phase and the $\sqrt{3} \times \sqrt{3}$ structure. For particles with a mass $M = 10\ m_{H_2}$ we find at low temperatures ($T < 16K$) a doubly peaked density distribution with a peak at the low density gas phase and one peak at a fluid density which is much higher than the monolayer density ($\rho > \rho_{\sqrt{3}}$). We conclude that for large particle masses the quantum delocalization is not very pronounced at low temperatures and so the repulsive interaction between the particles in the surface potential due to their effective "size" is reduced and smaller distances compared to the H_2 system are fre-

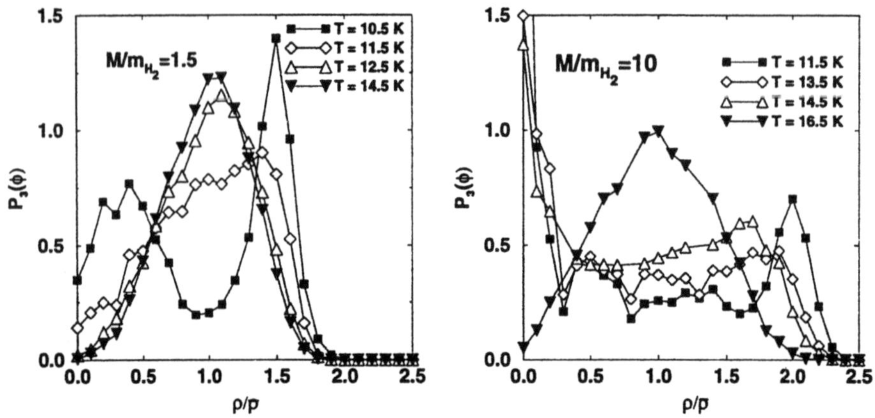

Figure 2: Density distributions in subsystems of size $L/S = 3$ for different temperatures (PIMC simulation with $P = 25$ and $N = 108$). Left panel: mass of the particles $M = 1.5m_{H_2}$; right panel: mass of the particles $M = 10m_{H_2}$;

quent resulting in a destabilization of the $\sqrt{3} \times \sqrt{3}$ structure and stabilizing a liquid phase instead. The results for the gas–liquid coexistence region for large masses are in good agreement with the critical point estimates for the (classical) two dimensional Lennard–Jones fluid [6] ($T_c = 17.25K$, $\rho_c = 0.94\bar{\rho}$). Further detailed studies of these effects are under current investigation [7].

Acknowledgements

Financial support from the DFG (DFG–Ni 259/8–1 for M.K. and Heisenberg Fellowship for P.N.) are gratefully acknowledged as well as the granting of computer time on the Cray-YMP (HLRZ Jülich and RHRK Kaiserslautern), VP 100 (RHRK Kaiserslautern).

References

[1] *Ordering in Two Dimensions*, edited by S. K. Sinha (Amsterdam, North-Holland, 1980).

[2] H. Freimuth, H. Wiechert, Surf. Sci. **178**, 716 (1986).

[3] X.Z. Ni, L.W. Bruch, Phys. Rev. B**33**, 4584 (1986).

[4] C.S. Murthy, K. Singer, M.L. Klein and I.R. McDonald, Mol. Phys. **41**, 1387 (1980); W.A. Steele, Surf. Sci. **36**, 317 (1973).

[5] K. Binder, Z. Phys. B**43**, 119 (1981); M. Rovere, D.W. Heermann and K. Binder, Europhys. Lett. **6**, 585 (1988).

[6] M. Rovere, P. Nielaba and K. Binder, Z. Phys. B **90**, 215 (1993).

[7] Martina Kreer, Ph.D. thesis (Mainz), in preparation.

The Path-Integral Monte Carlo Method for Rotational Degrees of Freedom

M.H. Müser

Department of Chemistry, Columbia University, New York, NY 10027, USA

Abstract. Very recent progress in the path-integral simulation of quantum mechanical rotational degrees of freedom (RDOF) is reviewed. Unlike translational degrees of freedom, where the free particle kernel can be obtained by a Gaussian integral, the kernel for RDOF is not analytically accessible. However, it can be obtained numerically to high precision on a fine grid so that all systematic errors can be kept controllably small. The formulae to generate the kernel for RDOF are given explicitly for one, two, and three-dimensional rotation. In particular, we discuss the consequences of confining the rotational state of a rigid body to one representation for the sign problem and the correlation time. A linear rotor in a Devonshire potential and solid methane are considered as examples.

1 The Path-Integral for RDOF

Path-integral Monte Carlo is a powerful method to study numerically the static low-temperature properties of clusters, liquids and solids. Sophisticated methods exist for the simulation of translational degrees of freedom [1, 2]. The computational treatment of quantum mechanical rotational degrees of freedom (RDOF), however, is rather tedious for the three important cases, the 1-d rotation (molecule confined to a surface), the 2-d rotation (linear molecule in three dimensional space) and the 3-d rotation (non-linear molecule in $d = 3$). The reason is the difficulty in determining analytically the free-particle kernel

$$K_f(i,j,\beta/P) = \langle i \mid \exp\left(-\beta\hat{T}_{\rm rot}/P\right) \mid j \rangle \tag{1}$$

in imaginary time as a function *exclusively* of orientations $\mid i \rangle$ and $\mid j \rangle$. Possible phase factors arising from nuclear spins will be discussed later.

For the 1-d rotation the so-called winding number n comes into play (giving rise to large correlation times) and $K_f(\varphi_i, \varphi_j, \beta/P)$ may be expressed as [3]:

$$K_f(\varphi_i,\varphi_j,\frac{\beta}{P}) = \sqrt{\frac{P}{2\pi\beta B_{\rm rot}}} \sum_{n=-\infty}^{\infty} \exp\left[-P(\varphi_i - \varphi_j + \pi n \delta_{iP})^2/\beta B_{\rm rot}\right] . \tag{2}$$

In the case of 2-d rotation a winding number does not exist due to different topologies of the one and 2-d unit sphere and it is not possible to find analytically an expression for K_f. An alternative is to integrate out the stiff bonds of the molecules. This procedure, however, implies the necessity of huge Trotter numbers, which in turn usually lead to large correlation times and large

Springer Proceedings in Physics, Volume 82
Computer Simulation Studies in Condensed-Matter Physics IX
Eds.: D. P. Landau, K.K. Mon, H. -B. Schüttler

statistical error bars. Hence one is disabled in investigating the systems at $k_B T \leq B_{\mathrm{rot}}$.

In real space the rotational state of a linear molecule can be represented by a normalized vector $\vec{n}_i$ indicating the orientation of the rotor at Trotter slice i. We thus need to compute

$$K_f\left(\vec{n}_i, \vec{n}_j, \beta/P\right) = < \vec{n}_i \mid \exp\left(-\beta B_{\mathrm{rot}} \vec{L}^2/P\hbar^2\right) \mid \vec{n}_j > \tag{3}$$

in order to develop a PIMC simulation. Throughout this proceeding we disregard the potential energy V because the $\mid \vec{n}_i\rangle$ are eigenstates of V. Hence the inclusion of V is straightforward.

Inserting the identity operator $\hat{1}$ built up of the spherical harmonics $\mid l\, m\rangle$ and applying the addition theorem of the spherical harmonics leads to [4]:

$$K_f\left(\vec{n}_i, \vec{n}_j, \frac{\beta}{P}\right) = \sum_l \frac{2l+1}{4\pi} P_l(\vec{n}_i\, \vec{n}_j) \exp\left[-\frac{\beta}{P} l(l+1) B_{\mathrm{rot}}\right], \tag{4}$$

with $P_l(\vec{n}_i\, \vec{n}_j)$ the Legendre polynomial of degree l.

It is obviously a bad idea to include l as a Monte Carlo variable in the way the winding number has been included as a random variable for the treatment of 1-d RDOF. $P_l(\vec{n}_i\, \vec{n}_j)$ can have negative values which is a necessary and sufficient condition to obtain the sign problem [5, 6]. In this particular case the sign-problem is severe even at high temperatures because high l values are thermally allowed and only small changes in the orientations induce a negative sign of the probability measure.

By performing numerically the sum over l in Eq. (4) it is easy to include spin statistics prior to the simulation. For heteronuclear diatomic molecules all l values have to be taken into account, while for homonuclear molecules the rotational state is confined to either even l or odd l. For this reason, H_2 exists either as p-H_2 or as o-H_2 respectively depending on the total nuclear spin.

In the case of 3-d RDOF a similar approach has been proposed recently in order to replace the fixed-axis approximation [7]. The state of a molecule at Trotter slice i may be described by its three Euler angles $\omega_i = \phi_i, \theta_i, \chi_i$. Again an appropriate $\hat{1}$ must be inserted into Eq. (1). The identity operator is conveniently built up of wave functions $\mid JM\tilde{K}\rangle = \sum_{K=-J}^{J} A_{\tilde{K}M}^{(JM)} \mid JMK\rangle$, being eigenstates of the kinetic energy operator [8]:

$$\hat{T}_{\mathrm{rot}} \mid JM\tilde{K}\rangle = E_{\tilde{K}}^{(JM)} \mid JM\tilde{K}\rangle. \tag{5}$$

Apart from a factor $\sqrt{(2J+1)/8\pi^2}$, the $\mid JMK\rangle$ are the elements of the $(2J+1)$-dimensional irreducible representation $\mathcal{D}_{MK}^{J}(\phi, \theta, \chi)$ of the operator $\hat{D}(\omega) = \exp(-i\hat{J}_z\phi)\exp(-i\hat{J}_y\theta)\exp(-i\hat{J}_z\chi)$. If we now express $\mid \omega_j\rangle$ in the body-fixed frame of the molecule at Trotter slice i, we arrive at [9]:

$$K_f(\omega_i, \omega_j; \frac{\beta}{P}) = \sum_{JM\tilde{K}} \frac{2J+1}{8\pi^2} d_{MM}^{(J)}(\tilde{\theta}_j) \mid A_{\tilde{K}M}^{(JM)} \mid^2 e^{iM(\tilde{\phi}_j + \tilde{\chi}_j) - \beta E_{\tilde{K}}^{(JM)}/P}, \tag{6}$$

where $\tilde{\phi}_j, \tilde{\theta}_j, \tilde{\chi}_j$ are the Euler angles of molecule j expressed in the molecule-fixed frame of molecule i and $d^{(J)}_{MM'}(\theta) = \langle JM \mid \exp(-i\hat{J}_y\theta) \mid JM'\rangle$.

In a first attempt to include spin statistic effects the rotational state of the top has to be confined to one representation only. Further restrictions may be necessary if the nuclear-spin state of the molecule is known in detail. Wave functions belonging to the representation μ are usually generated with the help of a projection operator $\hat{P}_\mu$ (see any textbook on group theory):

$$\hat{P}_\mu = \frac{d_\mu}{g} \sum_R \chi_R^{(\mu)*} \hat{D}(R), \tag{7}$$

where we sum over all symmetry operations (rotations) R of the molecule. If we apply $\hat{P}_\mu$ to the non-symmetrized rotational states in Eq. (6) the kernel $K_f^{(\mu)}$ belonging to the representation μ is obtained:

$$K_f^{(\mu)}(\omega_i, \omega_j; \frac{\beta}{P}) = \frac{d_\mu}{g} \sum_R \chi_R^{(\mu)*} K_f \left(0, \tilde{\omega}_j(R); \frac{\beta}{P}\right), \tag{8}$$

with $\mid \tilde{\omega}_j(R)\rangle = \hat{D}^{+}(\omega_i)\hat{D}(R) \mid \omega_j\rangle$.

Expectation values of observables such as internal energy and specific heat are obtained as usual by performing the derivatives of $\ln Z(\beta)$ with respect to the relevant variables, where

$$Z(\beta) = \int d\omega_1 \ldots \int d\omega_P \prod_{i=1}^{P} [K_f(\omega_i, \omega_{i+1}; \beta/P) \exp(-\beta V(\omega_i)/P)] \tag{9}$$

and $\omega_1 = \omega_{P+1}$.

2 The Role of Symmetrized Kernels in PIMC

Application of Eqs. (7) and (8) to linear molecules leads to the above mentioned (anti-) symmetrized kernels (o-H_2,p-H_2) or to non-symmetrized kernels (HD).

In Fig. 1(a) we show the effect of the (anti)-symmetrization on the kernel in imaginary time for the 2-d rotation. In the non-symmetrized and in the symmetrized case the kernel is non negative while the sign-problem is obtained in the case of the anti-symmetrized kernel. Thus we have to compute expectation values of an observable O according to [5]:

$$\langle O\rangle = \lim_{M\to\infty} \sum_{t=1}^{M} \sigma_t O_t / \sum_{t=1}^{M} \sigma_t, \tag{10}$$

where t enumerates the Monte Carlo steps and σ_t denotes the sign of the Boltzmann weight at Monte Carlo step t.

In Fig. 1(b) we show the average $\langle\sigma\rangle$ of the sign as a function of temperature for the free anti-symmetric rotor and for an anti-symmetric rotor in a Devonshire potential with an energy barrier 40 times B_{rot} [4]. At high temperatures

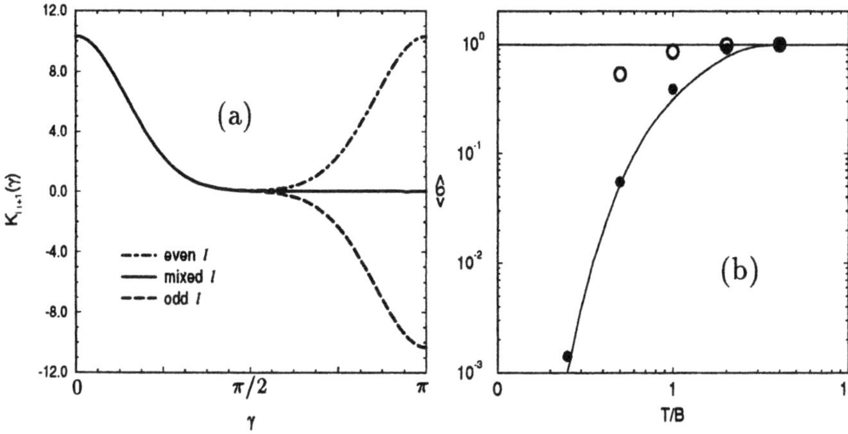

Fig. 1: (a) The kernel for $\beta B_{\rm rot}/P = 0.1$ and $\gamma = \cos(\vec{n}_i \vec{n}_{i+1})$. (b) $\langle\sigma\rangle$ as a function of temperature; full circles are numerical results for the free rotor, open circles correspond to a rotor in a Devonshire potential [4].

the sign is very close to one . At low temperatures this is no longer valid because the "orientational de Broglie wavelength" becomes larger and interesting quantum phenomena such as tunneling and exchange occur.

A similar behavior can be found for the kernel of non-linear molecules. The full kernel and the kernel corresponding to the identity representation are positive valued functions while other representations invoke the sign-problem.

There seem to be two major advantages in chosing the identity representation when molecular systems are investigated at low temperatures: (i) the correlation time for many observables are smaller and (ii) thermally excited states are suppressed such that "ground state properties" can be accessed even at relatively moderate temperatures. E.g., the rotational constant of CH_4 is $B_{\rm rot} \approx 7.54$ k_BK. The two smallest allowed J-values in the identity representation are $J = 0$ and $J = 3$. Thus the energy of the first excited state of the free molecule (in the identity representation) is $E_1 \approx 90$ k_BK in comparison to $E_1 \approx 15$ k_BK in case of the full kernel.

In Fig. 2(a) we show a relaxation function for the estimator of the kinetic energy of solid CH_4 investigated at $T = 16$ K with $P = 4$. The correlation time using the identity representation obviously is much smaller than in case of the full kernel. In Fig. 2(b) we show the internal energy of "classical" CH_4 and of "quantum mechanical" CH_4. A lowering of the transition temperature in the order of 40% is observed due to quantum delocalization [9].

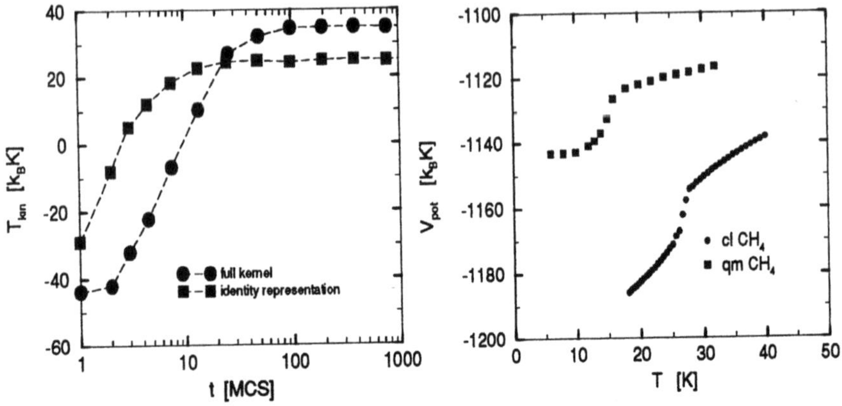

Fig. 2: (a) Relaxation of the kinetic energy estimator T_{kin} to its equilibrium value as a function of number of MC steps. (b) Potential energy V_{pot} as a function of temperature for "classical" and "quantum" CH_4.

3 Conclusions

We have discussed techniques we have developed to apply to path-integral Monte Carlo simulations. With the presented methods for the simulation of rotational degrees of freedom it is possible to perform PIMC simulations with the same efficiency as is the case for translational degrees of freedom. The basic idea is to evaluate numerically the free particle kernel prior to the simulation and to tabulate it on a fine grid such that all systematic errors are contollably small. We even suggest using such tables for the 1-d rotation. This will obviously reduce the correlation times at low temperatures because the winding number will no longer be a stochastic variable.

It is also possible to include spin statistics effects that confine the rotational wavefunction of the free molecule to one representation. However, only the full kernel and the kernel belonging to the identity representation are positive valued functions. For all other representations the sign problem is obtained. Confining the rotational state at low temperatures to the identity representation leads to much smaller correlation times in observables such as energy and imaginary time-correlation functions. With this restriction many thermally excited states are projected out such that it even might be possible to investigate ground state properties by PIMC.

The methods are conveniently applied to solids, clusters and liquids of light rigid molecules such as H_2,CH_4, NH_3, and H_2O at low temperatures.

Acknowlegments. The author thanks B.J. Berne for useful discussions as well as the Alexander von Humboldt Foundation for a Feodor Lynen Fellowship.

References

[1] *Quantum Simulation of Condensed Matter Phenomena*, Proceedings on an international Workshop, edited by J.D. Doll and J.E. Gubernatis (World Scientific, Singapore, 1990).

[2] K.E. Schmidt and D.M. Ceperley, in *Monte Carlo techniques for quantum fluids, solids and droplets*, Topics in Appl. Phys. Vol. 71, edited by K. Binder (Springer, Berlin, 1992).

[3] L.S. Schulman, *Techniques and Applications of Path Integrations* (John Wiley & Sons, New York, 1981).

[4] M.H. Müser, Molecular Simulation, in press.

[5] E.Y. Loh, Jr. and J.E. Gubernatis, R.T. Scalettar, S.R. White, D.J. Scalapino, and R.L. Sugar, Phys. Rev. B **41**, 9301 (1990).

[6] H. De Raedt and M. Frick, Phys. Rep. **231**, 107 (1993).

[7] R.A. Kuharski and P.J. Rossky, J. Chem. Phys. **82**, 5164 (1985).

[8] R.N. Zare, *Angular Momentum* (John Wiley & Sons, New York, 1988).

[9] M.H. Müser and B.J. Berne, *The path-integral Monte Carlo scheme for rigid tops* , (preprint).

Three-State Potts Model on the Quasiperiodic Octagonal Tiling

T. Boutry[1], *D. Ledue*[1], *and D.P. Landau*[2]

[1]LMA URA CNRS 808, Faculté des Sciences,
F-76821 Mont-Saint-Aignan Cédex, France
[2]Center for Simulational Physics, The University of Georgia,
Athens, GA 30602, USA

Abstract. The static critical behaviour of the three-state Potts model on the two-dimensional (2D) quasiperiodic octagonal tiling is studied by means of Monte Carlo simulations. Our investigation has been carried out using a finite-size scaling analysis of octagonal tilings with free boundary conditions. The estimated critical exponents α, γ and ν strongly suggest that the three-state Potts model on the octagonal tiling belongs to the same universality class as the three-state Potts model on 2D periodic lattices.

1. Introduction

The q-state Potts model [1] has been widely investigated for twenty years [2]. The nature of the phase transition in the two-dimensional (2D) q-state Potts model on periodic lattices has been known to be first order for $q > 4$ and second order for $q \leq 4$ [3,4]. Moreover, the transition temperature is known for all q as are the critical exponents for $q \leq 4$. More recently, a rather intriguing feature concerning a change in the nature of the transition for the 2D eight-state Potts model due to the addition of quenched, bond randomness has been reported [5,6]. This study has evidenced that the phase transition changes from first order to second order with 2D Ising exponents.

Unlike periodic lattices, studies about classical spin systems on 2D quasiperiodic tilings have been carried out only since 1988 [7], few years after the discovery of quasicrystals (in 1984) [8]. Actually, only the Ising model ($q = 2$) on 2D quasiperiodic tilings has been investigated. For the non-frustrated ferromagnetic Ising model on the Penrose tiling (5-fold symmetry) and on the octagonal tiling (8-fold symmetry) (Fig. 1), the critical exponents have been found to be the same as for 2D periodic lattices [7,9,10]. On the other hand, the estimated critical temperatures, $kT_c/J = 2.392\pm0.004$ and $kT_c/J = 2.401\pm0.005$ for the Penrose tiling [7,9], and $kT_c/J = 2.39\pm0.01$ for the octagonal tiling [10] are slightly higher than the critical temperature in the square lattice ($kT_c/J = 2.269$) while the mean coordination number is equal to 4 for all three lattices. In the same way, a study of a weakly frustrated ferromagnetic Ising model on the octagonal tiling has provided the same critical exponents [11].

In this paper, we investigate the static critical behaviour of the three-state Potts model on the quasiperiodic octagonal tiling. This work has been motivated by the wealth of the Potts model, as previously reminded, and by the unusual properties concerning long-range order in quasiperiodic tilings as mentionned below.

Springer Proceedings in Physics, Volume 82
Computer Simulation Studies in Condensed-Matter Physics IX
Eds.: D. P. Landau, K.K. Mon, H. -B. Schüttler

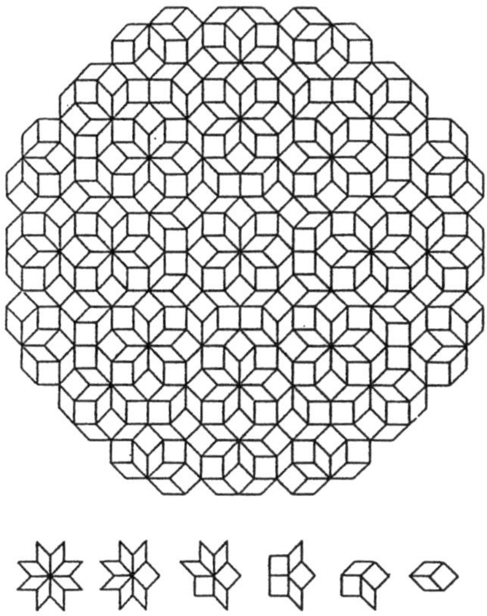

Fig. 1. The octagonal tiling (8-fold orientational symmetry) and its six local environments.

A quasiperiodic tiling is constructed by the infinite repetition in space of two or more distinct " unit cells ". The set of the vertices of such a tiling is a lattice that exhibits long-range quasiperiodic translational order and long-range orientational order [12], which can be in particular, 5-fold or 8-fold. Two quasiperiodic tilings which exhibit the same long-range orientational order are locally isomorphic if they can be made to overlap out to any finite distance by a finite translation. This definition implies that two quasiperiodic tilings which are locally isomorphic are equivalent on the average, that is, they exhibit identical macroscopic physical properties (transition temperature, critical exponents, ground states...) [12]. It can be demonstrated that all the octagonal tilings are locally isomorphic, then, the static critical behaviour of a system of spins located at the vertices of an octagonal tiling should not depend on the particular tiling which is considered.

The model, simulation techniques and finite-size scaling theory are presented in Section 2. Results are discussed in Section 3 and a summary is given in Section 4.

2. Background

The Hamiltonian of the q-state Potts model [1] is given by :

$$H = -J \sum_{<i,j>} \delta_{\sigma_i \sigma_j} \quad (J > 0)$$

where the spins σ_i, which are located at the vertices of the octagonal tiling, take on the values 1, ..., q and δ is the Kronecker delta function. The ferromagnetic interaction J is along the edges of the unit cells of the tiling (the sum goes over all next nearest-neighbour pairs in the tiling), so the mean coordination number is equal to 4. The order parameter M has been defined by $M = (qm-1)/(q-1)$, where $m = L^{-d} \max(N_1, N_2, ..., N_q)$ in which N_i is the number of spins in state i and $L^d = N$ is the total number of spins [6]. In this study, $d = 2$ and $q = 3$.

Numerical simulations were carried out on finite octagonal tilings of sizes N = 329, 689, 1433, 2929 and 5497 with free boundary conditions. The procedure is a Monte Carlo (MC) procedure based on the standard single spin-flip Metropolis algorithm [13,14]. At each temperature, the first 20 000 MC steps per spin were discarded before averaging over the next 180 000 MC steps per spin (N = 329) up to 480 000 MC steps per spin (N = 5497).

The static critical behaviour of an infinite system may be deduced from the size dependence of several thermodynamic quantities which, in the case of a second order transition, includes the infinite system critical exponents α, β, γ and ν [15-17]. In particular, the order of a phase transition, which occurs at T_c, may be determined from the scaling form of the maximum in the specific heat [6,18] :

$$C_{max}(L) \sim L^{\alpha/\nu}\, C(t\, L^{1/\nu}) \text{ for second order transitions,}$$

$$C_{max}(L) \sim L^{d} A(t\, L^{d}) \text{ for first order transitions,}$$

where $t = |T-T_c|/T_c$. For a second order transition, the critical exponent ν can be estimated from the logarithmic derivatives ϕ_n of any power of the order parameter : $\phi_n(T,L) = \dfrac{\partial \ln < |M|^n >}{\partial T^{-1}}$ [19]. The magnitude of the maximum in ϕ_n, varies with the system size, asymptotically, as $L^{1/\nu}$. In our analysis, we have only considered ϕ_1 and ϕ_2. The infinite system critical temperature T_c can be extracted from the location $T_c(L)$ of the maximum in ϕ_n, in the specific heat C, in the susceptibility χ and in the derivative of the order parameter $\partial M/\partial T^{-1}$:

$$kT_c(L)/J = kT_c/J + a\, L^{-1/\nu}$$

where a depends on the thermodynamic quantity [19,20]. The critical exponent γ can be estimated from the size dependence of the maximum value of the susceptibility [19] :

$$\chi_{max}(L) \sim L^{\gamma/\nu}\, \chi(t\, L^{1/\nu}).$$

The static critical exponents satisfy the hyperscaling relation $2-\alpha = d\nu = 2\beta + \gamma$ [21].

3. Results

For N = 329 up to N = 2929, two runs were performed and the final thermodynamic quantities have been obtained by averaging over the two runs. On the other hand, for N = 5497, the maximum value of the thermodynamic quantities is very difficult to

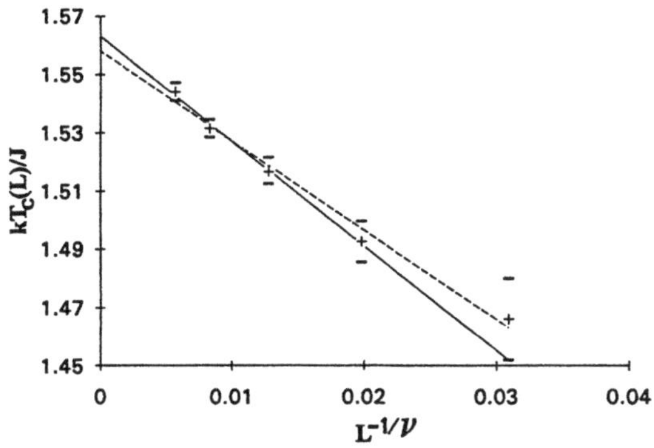

Fig. 2. Size dependence of the " effective transition temperature " for the specific heat (linear fit with $N_{min} = 329$: dashed line, linear fit with $N_{min} = 689$: continuous line).

determine for each single run because of a lack of data point near the " transition " (specially for ϕ_1 and ϕ_2). Therefore, for each thermodynamic quantity, we plotted on the same figure the temperature dependence for all runs (3 runs) in order to have more data points. Then, we were able to estimate quite accurately the maximum in the thermodynamic quantities for N = 5497.

For each plot of size dependence, three different linear fits were performed using system sizes $N_{min} \leq N \leq 5497$ with, respectively, $N_{min} = 329$, 689, or 1433. For each thermodynamic quantity, no significant deviation to the asymptotic linear regime has been noticed evidencing that the asymptotic finite-size scaling regime is already reached by N = 329. Therefore, where not mentionned, results which are given here correspond to the linear fit with $N_{min} = 329$.

In order to check the nature of the phase transition, we plotted the size dependence of the maximum value in the specific heat on a log-log scale. The slope of the linear fit is 0.43±0.01 which indicates that the transition is of second order with $\alpha/\nu = 0.43 \pm 0.01$.

The estimates of ν deduced from the size dependence of the maximum value of ϕ_1 and ϕ_2 are, respectively, 0.83±0.04 and 0.83±0.03. These estimates are in good agreement with the value for 2D periodic lattices ($\nu = 5/6$). Combining $\nu = 0.83 \pm 0.04$ with our estimate of α/ν, we deduced $\alpha = 0.36 \pm 0.02$ which is consistent with the 2D periodic value ($\alpha = 1/3$).

In order to determine the infinite octagonal tiling critical temperature, we plotted the size dependence of the " effective transition temperatures " $T_c(L)$ for the specific heat (Fig. 2), the susceptibility and the derivative of the order parameter. Our three estimates, which have been extracted by averaging over the linear fits with $N_{min} = 329$ and $N_{min} = 689$, are respectively : 1.560±0.005, 1.560±0.005 and 1.556±0.006. Combining these values, we obtain $kT_c/J = 1.559 \pm 0.006$ which is slightly higher than the critical point of the three-state Potts model on the square lattice ($kT_c/J = 1.492$). It should be noted that the ratio

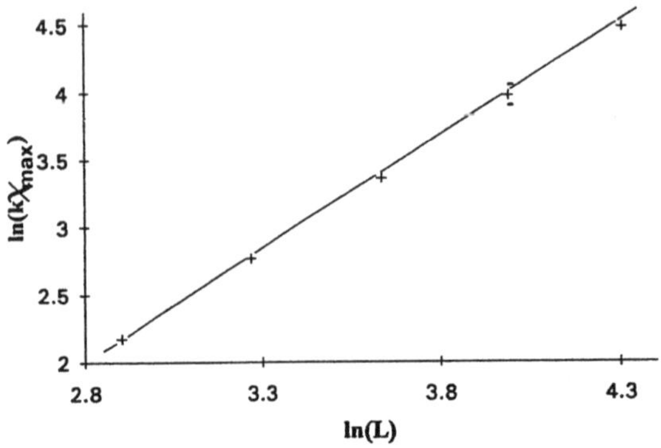

Fig. 3. Log-log plot of the size dependence of the maximum in the susceptibility (where not shown, error bars are smaller than the symbols).

$(kT_c/J)_{octagonal} / (kT_c/J)_{square}$ for the three-state Potts model is ≈ 1.045 which is roughly equal to the ratios for the Ising model : $(kT_c/J)_{octagonal} / (kT_c/J)_{square} \approx 1.053$ and $(kT_c/J)_{Penrose} / (kT_c/J)_{square} \approx 1.056$.

The plot of the size dependence of the maximum in the susceptibility on a log-log scale is shown on Figure 3. This plot provides $\gamma/\nu = 1.64\pm0.03$. Taking into account our estimate of ν, we found $\gamma = 1.37\pm0.09$ which is in reasonable agreement with the 2D periodic value ($\gamma = 13/9$).

Concerning the hyperscaling relation, our results lead to $2-\alpha = 1.64\pm0.02$ and $d\nu = 1.66\pm0.08$.

4. Conclusion

Our investigation has shown that the finite-size scaling theory describes the size behaviour of the three-state Potts model on finite octagonal tilings with free boundary conditions. Our estimates of the critical exponents of the three-state Potts model on the octagonal tiling are clearly consistent with the 2D periodic critical exponents. This strongly suggests that the three-state Potts model on the octagonal tiling belongs to the same universality class as the three-state Potts model on 2D periodic lattices. It should be noted that, as for the Ising model [7,9-11], the critical point in the octagonal tiling has been found slightly higher than the critical point in the square lattice confirming that tendency to ferromagnetic ordering in 2D quasiperiodic tilings is higher than in 2D periodic lattices.

Numerical simulations are still in progress in order to improve the accuracy of the estimates of the critical exponents.

Acknowledgments

The authors wish to thank the Centre de Ressources Informatiques de HAute-Normandie (CRIHAN) for providing computer facilities.

References

1. F. Y. Wu, Rev. Mod. Phys. **54**, 235 (1982)
2. K. Binder, J. Stat. Phys. **24**, 69 (1981)
3. R. J. Baxter, J. Phys. C **6** : L445 (1973)
4. R. J. Baxter, H. N. V. Temperley and S. E. Ashley, Proc. R. Soc. London A **358**, 535 (1978)
5. S. Chen, A. M. Ferrenberg and D. P. Landau, Phys. Rev. Lett. **69**, 1213 (1992)
6. S. Chen, A. M. Ferrenberg and D. P. Landau, Phys. Rev. E **52**, 1377 (1995)
7. Y. Okabe and K. Niizeki, J. Phys. Soc. Jpn **57**, 16 (1988)
8. D. Schechtman, I. Blech, D. Gratias and J. W. Cahn, Phys. Rev. Lett. **53**, 1951 (1984)
9. E. S. Sorensen, M. V. Jaric and M. Ronchetti, Phys. Rev. B **44**, 9271 (1991)
10. D. Ledue, D. P. Landau and J. Teillet, Phys. Rev. B **51**, 12523 (1995)
11. D. Ledue, Phys. Rev. B **53** (1996) (to be published)
12. D. Levine and P. J. Steinhardt, Phys. Rev. B **34**, 596 (1986)
13. S. Kirkpatrick, C. D. Gelatt Jr. and M. P. Vecchi, Science **220**, 671 (1983)
14. N. Metropolis, A. E. Rosenbluth, M. N. Rosenbluth, A. H. Teller and E. Teller, J. Chem. Phys. **21**, 1087 (1953)
15. M. E. Fisher, " Critical Phenomena ", edited by M. S. Green (Academic, New-York, 1971)
16. M. E. Fisher and M. N. Barber, Phys. Rev. Lett. **28**, 1516 (1972)
17. D. P. Landau, Phys. Rev. B **14**, 255 (1976)
18. M. S. S. Challa, D. P. Landau and K. Binder, Phys. Rev. B **34**, 1841 (1986)
19. A. M. Ferrenberg and D. P. Landau, Phys. Rev. B **44**, 3249 (1991)
20. K. Chen, A. M. Ferrenberg and D. P. Landau, Phys. Rev. B **48**, 3249 (1993)
21. V. Privman, P. C. Hohenberg and A. Aharony, " Phase Transitions and Critical Phenomena ", edited by C. Domb and J. L. Lebowitz (Academic, New-York, 1991), Vol. 14, p. 1

Application of the Exchange Monte Carlo Method to the 3D $\pm J$ Ising Spin Glass

K. Hukushima[1], *H. Takayama*[2], *and K. Nemoto*[3]

[1]Institute of Physics, University of Tsukuba, 1-1-1 Tennoudai, Tsukuba 305, Japan
[2]Institute for Solid State Physics, University of Tokyo, 7-22-1 Roppongi, Minatoku, Tokyo 106, Japan
[3]Department of Physics, Hokkaido University, Kita-ku, Sapporo 060, Japan

Abstract. Recently we have proposed the exchange Monte Carlo method for simulating hardly-relaxing systems. Its further details, such as behavior of the time correlation function, are investigated by applying it to the 3D $\pm J$ Ising spin glass.

1. Introduction

The low temperature phase of spin glasses and other complex systems generally have numerous local minima which are separated to each other by energy barriers. The characteristic time in which the system escapes from a local minimum increases rapidly as temperature decreases. This situation causes "hardly-relaxing" problem in using conventional Monte Carlo (MC) simulations based on a local updating. In order to overcome this difficulty we have recently proposed a new algorithms which we call the exchange MC method [1,2]. It is one kind of the extended ensemble methods, and can be regarded as an parallelized version of the simulated tempering [3], though the extended ensemble used is very different.

In the present report we briefly review our exchange MC method, and further investigate its details by applying it to the 3D $\pm J$ Ising spin glass. To characterize the time scale of relaxation in our method, we have observed the ergodicity time, which is defined as the average MC steps (MCS) per one travel over the whole temperature range, and the relaxation time of a modified autocorrelation function which exhibits an exponential decay. The two times are comparable with each other and take moderate values even at low temperatures below T_c, the spin-glass (SG) transition temperature.

2. Exchange MC method

In our method we treat a compound system consisting of non-interacting M replicas of the system concerned (whose Hamiltonian is specified as $\mathcal{H}(X)$). Each replica is in contact with its own heat bath having different inverse temperature β_m (for convenience we take $\beta_m < \beta_{m+1}$). A state of this ex-

Springer Proceedings in Physics, Volume 82
Computer Simulation Studies in Condensed-Matter Physics IX
Eds.: D. P. Landau, K.K. Mon, H. -B. Schüttler

tended ensemble is specified by $\{X\} = \{X_1, X_2, \cdots, X_M\}$, and its partition function $\mathcal{Z}$ is simply given by $\mathcal{Z} = \prod_{m=1}^{M} Z(\beta_m)$, where $Z(\beta_m)$ is that of the m-th replica. For a set of temperature $\{\beta\}$ given, the probability distribution of finding $\{X\}$ becomes $P(\{X,\beta\}) = \prod_m^M P_{\rm eq}(X_m, \beta_m)$, where $P_{\rm eq}(X,\beta) = Z^{-1}(\beta)\exp(-\beta\mathcal{H}(\mathrm{X}))$.

In constructing a Markov process for the exchange MC method we introduce a transition matrix $W(X, \beta_m|X', \beta_n)$ which is a probability of exchanging configurations (or exchanging temperatures) of the n-th and m-th replicas. In order that the system remains at equilibrium, it is sufficient to impose the detailed balance on the transition matrix which is specified by

$$\frac{W(X, \beta_m|X', \beta_n)}{W(X', \beta_m|X, \beta_n)} = \exp(-\Delta), \tag{1}$$

where

$$\Delta = (\beta_n - \beta_m)(\mathcal{H}(X) - \mathcal{H}(X')). \tag{2}$$

An explicit form of $W(X, \beta_m|X', \beta_n)$ is easily written down if one adopts, for example, the Metropolis method.

For the actual MC procedure, the following two steps are performed alternately: 1) Each replica is simulated *simultaneously* and *independently* as canonical ensemble for a few MCS by using a standard MC method. 2) Exchange of two configurations X_m and X_{m+1}, is tried and accepted with the probability $W(X_m, \beta_m|X_{m+1}, \beta_{m+1})$.

Here we restrict the replica-exchange to the case $n = m + 1$. The quantity Δ, which characterizes the probability of exchanging β_m and $\beta_{m+1} = \beta_m + \delta$, is estimated to order δ^2 as

$$\Delta = \delta(\mathcal{H}(X_{m+1}) - \mathcal{H}(X_m)) \sim \delta^2 \frac{\mathrm{d}}{\mathrm{d}\beta} E, \tag{3}$$

where the instantaneous value of the energy $\mathcal{H}$ is approximated by the thermal expectation value E. Since the energy E is an extensive variable, δ should be of order of $N^{-1/2}$ to satisfy the condition that $\Delta \sim O(1)$.

Another restriction in choosing a set $\{\beta\}$ is as follows. The highest temperature $1/\beta_1$ is high enough where there exists only one minimum in the free-energy structure and relaxation (correlation) time is very short, otherwise the system would not completely forget where it was trapped before even if it visits to the highest temperature. On the other hand, the lowest temperature $1/\beta_M$ is somewhere in the low temperature phase whose properties we are interested in. Thus the number of temperatures M we have to simulate is at least of the order of $\sqrt{N}$. Once we have appropriately chosen a set $\{\beta\}$ there remains no other parameters to be determined beforehand.

3. Application to the 3D $\pm J$ Ising spin glass

By means of the exchange MC method we have simulated the model spin-glass (with equal numbers of $+J$ and $-J$ bonds) on cubic lattices of linear size $L = 6$, 8, 12 and 16 with the periodic boundary condition.

The local updating have been performed with the multi-spin coding technique which simulates 32 different physical systems at once. This fixes the total number of temperatures to be $M = 32$ in all lattice sizes. The temperatures set $\{\beta_m\}$ is determined so as to obtain nearly equal acceptance probabilities for replica exchange trials at all β_m. This is done for a typical sample with $L = 12$. The obtained set in the range $0.86 \leq T/J \leq 2.39$ is used for all samples and sizes. The resultant acceptance ratios for various sizes are found to be as large as several tenths and not to depend on the bond realization so sensitively.

With the chosen set $\{\beta_m\}$, we investigate the ergodicity time $\tau_{\rm E}$ mentioned in Sec. 1. Its dependence on the lattice size is shown in Fig. 1, with that of the multi-canonical method by Berg et al [4]. In order check whether, within the time-scale of $\tau_{\rm E}$, the system really forgets where it was trapped, we study the autocorrelation function $q(t, \beta_m^0)$ of the m-th replica averaged along trajectory in the temperature space evolving from the initial temperature β_m^0. The function $q(t, \beta)$ is expected to include the slowest relaxation mode of the present method. In Fig. 2 we show $q(t, \beta)$ at various temperatures for $L = 12$ averaged over 10 samples. From the slope of the straight line (exponential decay) we evaluate the relaxation time τ of $q(t, \beta)$. The obtained τ exhibits a crossover feature at around $T_{\rm cr} \sim 1.5J$; at $T > T_{\rm cr}$ τ behaves like the relaxation time obtained by the conventional MC process alone, while $T < T_{\rm cr}$ τ become nearly saturated and comparable with τ_E.

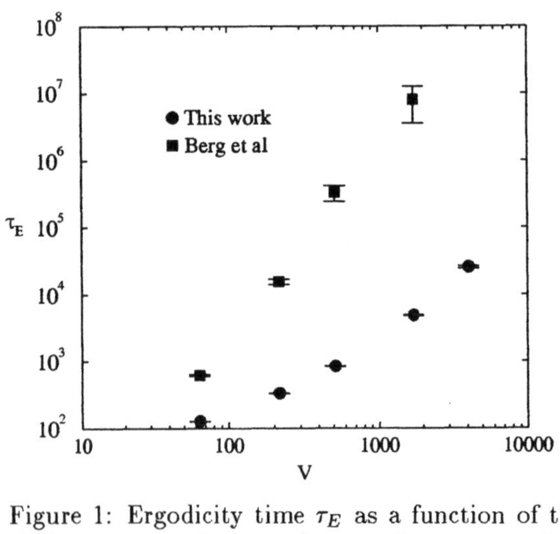

Figure 1: Ergodicity time τ_E as a function of the system size $V = L^3$. The circles correspond to our data, and squares to the multi-canonical method.

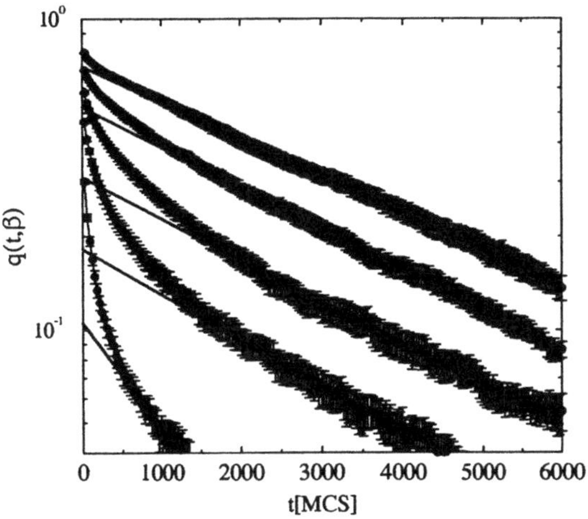

Figure 2: Autocorrelation function $q(t,\beta)$ for various temperatures. $1/\beta = T = 0.862$, 1.088, 1.256, 1.399, and 1.609 (top to down).

4. Discussion

Thus far we applied the exchange MC method to the 3D $\pm J$ Ising spin glass, and found that the system really traverses over wide temperature space and the largest relaxation time in this dynamics is given by the ergodicity time, which is much smaller than the conventional one. Consequently the order parameter distribution $P(q)$, which is symmetric in q, can be obtained below T_c up to $L = 16$ within much shorter MCS than that by the conventional MC method. Further details of the analysis will be reported elsewhere [2].

Our initial motivation of the present research was to establish our new method and then to make full use of it to re-examine nature of the phase transition of the same SG model, which have been recently redebated [5,6]. We also reported our preliminary results in [1], which agree qualitatively with those in [6]. We are now extending the analysis to include systems of a larger size ($L = 24$ with $M = 64$) and to increase statistical accuracy, i.e., the number of samples to be averaged. At the moment, however, our data are not enough to extract a definite answer to the problem.

Acknowledgments: Numerical calculations were mainly performed on Fujitsu VPP500 at the Institute for Solid State Physics, University of Tokyo. One of the authors (K.H) was supported by Fellowships of the Japan Society for the Promotion of Science for Japanese Junior Scientists.

References

1. K. Hukushima, H. Takayama and K. Nemoto, to appear in Intern. J. Mod. Phys. C (Proc. of the US-Japan Bilateral Seminar on "*New Trends in Computer Simulation of Spin Systems*").
2. K. Hukushima and K. Nemoto, preprint.
3. E. Marinari and G. Parisi: Europhys. Lett. **19**, 451 (1992).
4. B. A. Berg, U. Hansmann and T. Celik: Phys. Rev. B **50**, 16444 (1994).
5. E. Marinari, G. Parisi and Ritort: J. Phys. A **27**, 2687 (1994).
6. N. Kawashima and P.A. Young: to appear in Phys. Rev. B.

Large Finite-Size Effects of Discrete Systems in Microcanonical Ensemble Monte Carlo Simulations

J.R. Ray and C. Freléchoz

Department of Physics and Astronomy, Kinard Laboratory of Physics, Clemson University, Clemson, SC 29634-1911, USA

Abstract. We describe a microcanonical ensemble theory with a rigorous statistical mechanics basis which can be used for exact analytic calculations or Monte Carlo simulations of discrete systems. As an example we present results of simulations of a two dimensional Ising model on a square lattice in both the canonical and the microcanonical ensemble. The specific heat shows a difference of 16% for a 30x30 Ising model and 10% for a 60x60 Ising model system in zero magnetic field in 2-dimensions near the maximum in the specific heats.

1. Introduction

In earlier work [1] we discussed a microcanonical ensemble Monte Carlo method that can be used for simulations of systems with a continuous potential. Simulations of discrete spin systems using Creutz's demon algorithm [2] have been presented [3, 4] but since this method does not appear to have a rigorous basis in equilibrium statistical mechanics it is difficult to know how to interpret the results of such simulations. Here we show that the method presented in [1] can also be used for discrete systems, e.g. spin, lattice gas, or alloy type models. The method furnishes a way to carry out microcanonical ensemble simulations of these systems that has a rigorous basis in statistical mechanics. For simplicity of presentation we shall specialize the Hamiltonian to a Ising spin model in d-dimensions interacting with a magnetic field h. The Hamiltonian for the system can be written in the form

$$H(s,p)=\sum_{a=1}^{N}\sum_{j=1}^{d}\frac{p_{aj}^{2}}{2m}-J\sum_{<ab>}s_a s_b-h\sum_{a}^{N}s_a=K(p)+U(s) \tag{1}$$

where p_{aj} is the j component of the momentum of the a-th particle, s_a which is equal to +1 or -1 is the spin of the a-th particle, $K(p)$ is the kinetic energy of the system $U(s)$ is the potential energy of the system and $<ab>$ indicates a summation over nearest neighbors on the square lattice we employ. Usually we do not consider a discrete system, like the Ising model, to have a kinetic energy but we can associate the spin with particles of very large mass m so

Springer Proceedings in Physics, Volume 82
Computer Simulation Studies in Condensed-Matter Physics IX
Eds.: D. P. Landau, K.K. Mon, H. -B. Schüttler

there is no spatial motion. It seems clear that the real systems that are being modeled by the Ising interactions do in fact have mass but the spatial motion is not important for the magnetic properties of the system; letting the mass in Eq. (1) become arbitrarily large is a physical way to build in this property. For the canonical ensemble (system in contact with a thermal bath at temperature T) we have the equilibrium probability distribution W_C

$$W_c(s,p)=Ce^{-H(s,p)/(k_BT)} \quad , \tag{2}$$

where C is a normalization constant. Integrating over the momenta of the particles we arrive at the configuration probability distribution of the canonical ensemble

$$W_c(s)=C'e^{-U(s)/(k_BT)} \quad , \tag{3}$$

where C' is another normalization constant. Equation (3) shows that the momenta have no effect on the equilibrium properties of the system in the canonical ensemble and therefore all of the previous work which describes the Ising model using Eq. (3) are available. The configuration specific heat c_h of the system in the canonical ensemble can be determined from the potential energy fluctuations

$$c_h=\frac{1}{k_BT^2}\left(\langle U^2\rangle-\langle U\rangle^2\right) \quad . \tag{4}$$

For the microcanonical ensemble (isolated system in equilibrium) we have the probability distribution $W_m(s,p)$ which is constant on the energy shell $H(s,p) = E$,

$$W_m(s,p)=D\delta(E-H(s,p)) \quad , \tag{5}$$

where D is a normalization constant and δ represents the Dirac delta function. By integrating Eq. (5) over the momenta we obtain the configuration probability density for the microcanonical ensemble

$$Wm(s)=D'(E-U(s))^{(dN/2)-1}\Theta(E-U(s)) \quad , \tag{6}$$

where d is the spatial dimension of the system, $\Theta(x)$ is the unit step function which is 1 for $x>0$ and zero otherwise, and D' is another constant; the Θ function arises because the kinetic energy of the system $K=E-U$ is positive.

The temperature of the system is defined in terms of the average value of the kinetic energy $<K>=<E-U>$, using the probability density Eq. (6), $dNk_BT/2=\langle K\rangle$. If we differentiate this latter equation with respect to E then we obtain the fluctuation expression for the specific heat of the system in the microcanonical ensemble

$$\frac{d}{2c_h^t}=\frac{dN}{2}-(\frac{dN}{2}-1)\langle K\rangle\left\langle\frac{1}{K}\right\rangle \ , \qquad (7)$$

where c_h^t is the specific heat in the microcanonical ensemble. We determine the configuration specific heat by subtracting the ideal gas contribution $c_h= c_h^t -d/2$. The inverse kinetic energy term arises from the form of the probability distribution, Eq. (6). Although K appears in Eq. (7) we recall that K is a function of s through the relation $K(s)=E - U(s)$ and therefore the averages in Eq. (7) may be evaluated using Eq. (6). More fluctuation formulas in the microcanonical ensemble for this system are presented in [5].

2. Results and Discussion

Using the probability distributions for the canonical ensemble, Eq. (3), and microcanonical ensemble, Eq. (6), we carried out Monte Carlo simulations on a 2-dimensional square lattice Ising model with periodic boundary conditions in zero magnetic field. Simulations were carried out at nine temperatures in the canonical ensemble on a 30x30 Ising model. After equilibration we determined the average energy using the connection $<E> =dNk_BT/2+<U>$, where T is the assigned temperature and $<U>$ is determined by using the Metropolis Monte Carlo method with Eq. (3) as the importance function. Using this value for the energy in the microcanonical ensemble we can calculate the temperature using $dNk_BT/2=\langle E-U\rangle$ where we evaluate $<E-U>$ in the microcanonical ensemble using the Metropolis Monte Carlo procedure with Eq. (6). In Fig. 1 we show the energy, temperature pairs for both the canonical and microcanonical ensembles; the canonical ensemble values are shown by the squares and the full line while the microcanonical ensemble values are shown by the diamonds and the dashed line. As can be seen in Fig. 1. there is close agreement between the E versus T relation as calculated in the two ensembles. This implies that either ensemble may be used to study the system. In Fig. 2 we show the specific heats as determined in both ensembles using Eq. (4) for the canonical ensemble and Eq. (7) for the microcanonical ensemble. The two specific heats show close agreement except near the maximum, T≈ 2.30 where c(microcanonical)=2.167±0.012, while c(canonical)=1.871±0.004 for a 16% difference. Note that the two specific heats that are being compared are the same quantity; the difference is not like the

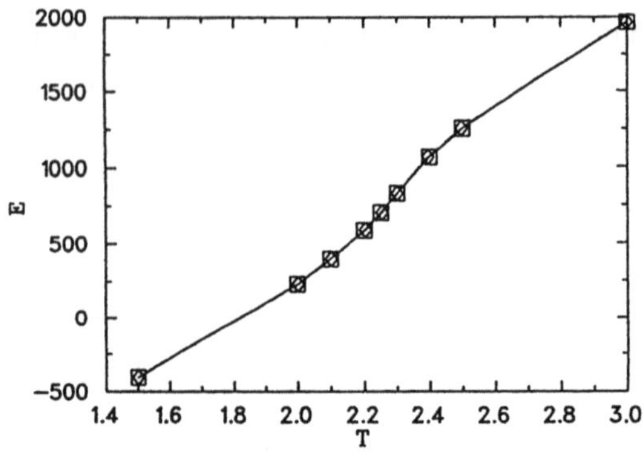

Figure 1. The energy versus temperature in the two ensembles for a 30x30 Ising model in 2-dimension with periodic boundary conditions and zero magnetic field. The squares and the full line are for the canonical ensemble while the diamonds and the dashed line are for the microcanonical ensemble. The lines are not distinguishable at this scale.

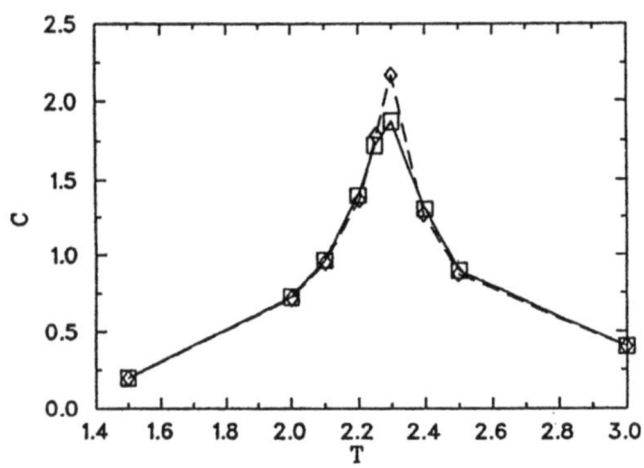

Figure 2. The specific heats in the two ensembles using the same convention as in Fig 1 for the symbols.

difference between the constant volume and constant pressure specific heats for a fluid but like the difference between the constant pressure specific heats calculated in two different ensembles! This large ensemble difference in specific heats is due to incipient phase transformation in the model. Note that in the thermodynamic limit (Onsager's solution) the specific heat is divergent at the transition temperature. The underlying phase transformation in the system in the thermodynamic limit magnifies the ensemble difference in these two ensembles for finite N giving rise to the 16% difference in the specific heats shown in Fig. 2. In Fig. 3 we show the

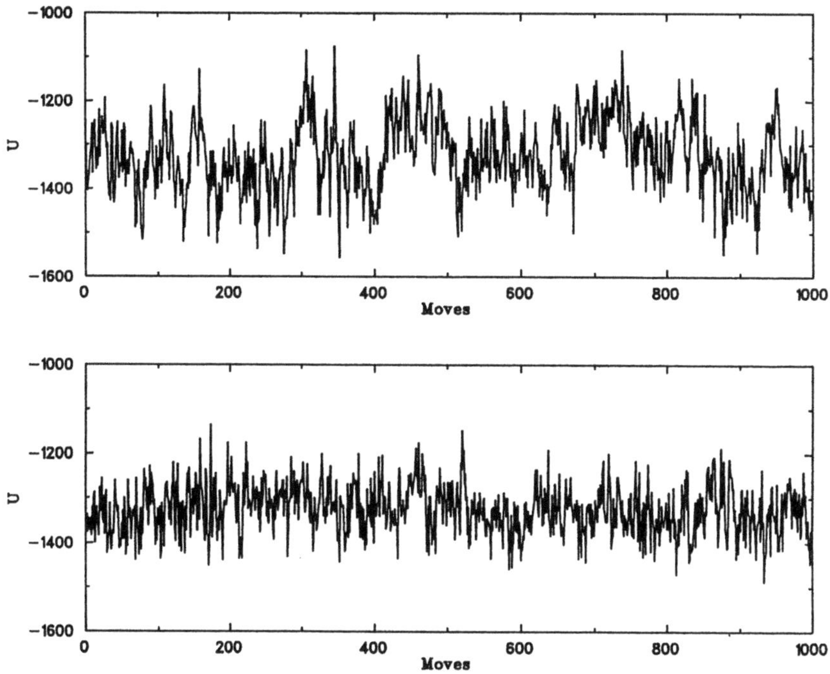

Figure 3. The upper panel shows the potential energy each move in a 1000 move canonical ensemble simulation at a temperature of 2.25, while the lower panel shows the microcanonical ensemble values for the energy corresponding to the temperature 2.25, namely 703. Note the qualitative difference in appearance of the fluctuations.

variation of U(s) over a 1000 move Monte Carlo run in the two ensembles for the 30x30 system. There is an interesting qualitative difference between the probability distributions in the two ensembles as shown in this figure. Finally in Fig. 4 we show the specific heats for a 60x60 spin system in the two ensembles in the vicinity of the maximum. There again is close agreement in the specific heats (not shown in this figure) except in the vicinity of the phase transformation where in this case the difference has narrowed to 10%. For the 60x60 lattice at this temperature the microcanonical and canonical specific heats are 2.464 and 2.226 respectively, hence, we see the expected increase in the value of the specific heat as well as the narrowing of the difference of the specific heats in the two ensembles for the larger system; from 0.30 to 0.24 or 16% to 10% for the 30x30 and 60x60 systems respectively.

We have generalized the microcanonical ensemble theory presented in [1] so that it applies to discrete systems. The method has a rigorous basis in

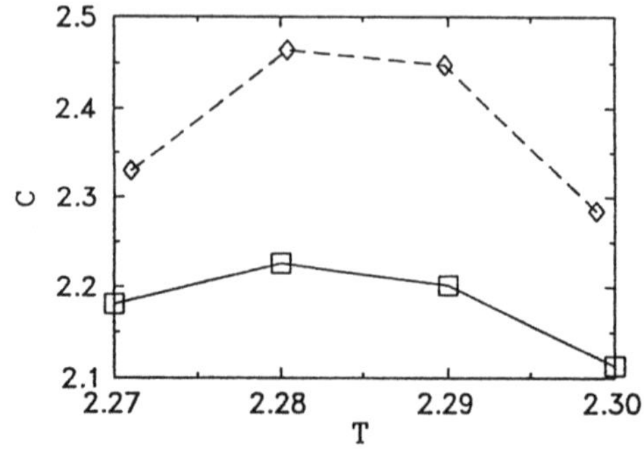

Figure 4. The specific heats for a 60x60 lattice with periodic boundary conditions near the maximum in the specific heats. The squares and full line are for the canonical ensemble while the diamonds and dashed lines are for the microcanonical ensemble.

equilibrium statistical mechanics and is just as easy to employ as the usual canonical ensemble procedure in Monte Carlo simulations. We have also shown that there are large ensemble differences between the canonical and microcanonical ensemble for finite lattice spin systems in the vicinity of the Onsager phase transformation.

References

1. J. R. Ray, Phys. Rev. A, 44, 4061 (1991).
2. M. Creutz, Phys. Rev. Lett. 50, 1411 (1983).
3. H. Gould and J. Tobochnik, Computer Simulation Methods, Part 2 (Addison-Wesley, Reading MA 1987).
[4]. D. W. Heermann, Computer Simulation Methods, 2nd ed. (Springer-Verlag, NY, 1990), and references contained therein.
[5] J. R. Ray and C. Freléchoz, Phys. Rev. E, March 1996 to appear.

Surface Critical Exponents of Linear Polymers on a Square Lattice: The Universality Classes of the θ and θ' Points

I. Chang[1,2] *and H. Meirovitch*

[1]Department of Physics, Pusan National University, Pusan 609-735, Korea

[2]Supercomputer Computations Research Institute, Florida State University, Tallahassee, FL 32306-4052, USA

Abstract : Using the scanning simulation method we study extensively the θ transition for self-avoiding walks, terminally attached to an adsorbing linear boundary on a square lattice, with an attraction energy for non-bonded nearest-neighbor monomers. The chains are simulated both at the special and ordinary points. Our critical exponents differ significantly from the exact values of Vanderzande, Stella, and Seno (VSS) [Phys. Rev. Lett. **67**, 2757 (1991)] for the θ' model. Thus their conjecture, that the θ and θ' points belong to the same universality class, is not supported.

I. Introduction

The collapse of polymers at the Flory θ-point[1, 2] and their adsorption on a surface are fundamental phenomena in polymer physics with a wide range of industrial applications[3] and biological importance (e.g. protein folding[4]). From the theoretical point of view, a great deal of progress has been achieved in recent years in two dimensions (2D), mainly due to the advent of Coulomb-gas techniques[5] and conformal invariance.[6] Duplantier and Saleur (DS)[7] proposed the exact tricritical exponents; the shape exponent $\nu = \frac{4}{7}$, the partition function exponent $\gamma = \frac{8}{7}$ and the crossover exponent $\phi = \frac{3}{7}$ of a linear polymers in $2D$. They also obtained the free energy surface exponents for a tricritical polymer that is terminally attached to a non-adsorbing impenetrable boundary (the ordinary point), $\gamma_1 = \gamma$ and $\gamma_{11} = \nu$. These exponents have been derived for a special model of SAWs on a hexagonal lattice with randomly forbidden hexagons, and consists, in addition to the nn attractions, also of a special subset of the next nearest neighbor attractions.[8 – 10] Therefore, instead of describing the usual θ-point, it might describe a multicritical θ'-point[11]. However, the strong discrepancy with the numerical results was found for the tricritical surface exponents of SAWs[12 – 15] ,where the numerical values are dramatically smaller than those of DS, i.e. $\gamma_1 \sim 0.6$ and $\gamma_{11} \sim -0.50$, vs. ~ 1.143 and ~ 0.571, respectively.

Recently Vanderzande, Seno and Stella (VSS)[16] have shown, that within the framework of the θ' model, the DS surface exponents $\gamma_1 = \frac{8}{7}$ and $\gamma_{11} = \frac{4}{7}$

Springer Proceedings in Physics, Volume 82
Computer Simulation Studies in Condensed-Matter Physics IX
Eds.: D. P. Landau, K.K. Mon, H. -B. Schüttler

are not related to the ordinary point but to the special one, i.e. the multicritical point at which adsorption transition of a θ' chain occurs and that the corresponding surface crossover exponent is $\phi_s = \frac{8}{21}$. They also derive $\gamma_1(\theta') = \frac{4}{7}$ at the ordinary point, which is compatible with numerical data for SAWs.[12, 15, 16] Thus, they have conjectured that the θ and θ' points are in the same universality class. In this paper we examine this conjecture by calculating the critical exponents of SAWs on a square lattice both at the special and ordinary points; We use the scanning simulation method which has been found very efficient to handle such problems.[13, 17, 18]

II. Model and Numerical Calculation

The SAWs, consisting of N steps, starts from the origin located on an adsorbing impenetrable linear boundary on the square lattice. An attractive energy ϵ_b is defined for two nn non-bonded monomers and an attractive surface energy ϵ_s is assigned to each bond on the surface. Three different microscopic energies can be defined for configuration i: The bulk energy $E_i^b = \epsilon_b m_i^b$, where m_i^b is the number of nn pairs of monomers. The surface energy, $E_i^s = \epsilon_s m_i^s$, where m_i^s is the number of bonds on the surface for a SAW, and the total energy, $E_i = E_i^b + E_i^s$. Two partition functions Z_α are defined $Z_\alpha = \sum_i{}^\alpha \exp(-E_i/k_B T)$ where k_B is the Boltzmann constant, T is the absolute temperature, $\alpha = 1$ denotes the usual partition function, while $\alpha = 11$ means that the summation is carried out only over the subgroup of chains which also end on the surface. The Boltzmann probability is $P_i^B = \exp(-E_i/k_B T)/Z_1$. Thus, the average surface energy E_s per ϵ_s reads $E_s = \epsilon_s^{-1} \sum_i P_i^B E_i^s$. It proves convenient to define the bulk and surface reciprocal temperatures, $K_b = -\epsilon_b/k_B T$ and $K_s = -\epsilon_s/k_B T$, where their critical values are denoted by K_b^* (θ point) and K_s^* (special point) respectively.

With the scanning method[19] a SAW is generated step-by-step by scanning all the possible chain continuations in b future steps. Since not the whole future is scanned, the chain can be trapped in a cul-de-sac; in this case it is discarded and a new chain is started. Therefore, from n_0 attempted chains only n will survive. Also, SAW i is not constructed according to its Boltzmann probability, P_i^B, but with a biased probability $P_i(b)$, which approachs P_i^B as b is increased. The bias can be removed by a procedure due to to Schmidt,[20] in which effectively smaller sample (the accepted Boltzmann sample) is extracted from the biased one. Thus, n_{accept} the number of *different* chains accepted to the unbiased sample serves as the effective sample size for the importance sampling results.[21]

III. Results and Discussion

The search for the special point K_s^* is based on the fact that with the scanning method results at *many* different temperatures can be obtained from a *single* sample simulated at a given temperature. We generated with the scanning method SAWs of $N = 200$ with $b = 3$. The number of attempted chains is relatively large, $n_0 \sim 40 \times 10^6$; however, the number of accepted chains $n_{\rm accept}$ becomes significantly lower as N increases; for SAWs of N=200 it is $\sim 0.9 \times 10^6$. The sample for SAWs was calculated at (K_b^*, K_s)=(0.658,0.810) (K_b^* was taken from Ref. 13) and results were obtained for the 34 temperatures, K_s=0.760, 0.763 ... 0.859 and K_b=0.654 and 0.662, which define the error bars for K_b^*. For the ordinary point we generated SAWs of $N = 250$ with $b = 5$ at K_b=0.628, 0.630 ... 0.676 and $n_0 = 186 \times 10^6$, $n_{\rm accept}(N = 200) \sim 3 \times 10^6$. In order to investigate the dependence of various properties on N, their importance sampling values were calculated and accumulated for the partial chains of lengths 10, 20, ... The surface critical temperature K_s^* was determined from the behavior of the surface energy E_s at (K_b^*, K_s^*),[22, 23]

$$E_s \sim N^{\phi_s} .$$

Therefore, at this point one expects, $E_s(2N)/E_s(N) = 2^{\phi_s}$ (if corrections to scaling are ignored). These ratios for N=10,20,... can be plotted as a function of K_s where the intersection point defines both K_s^* and ϕ_s.[24] A similar analysis was also made for estimating K_b^*. A careful analysis (which gives a larger weight to the results of the longer chains) leads to the following values,

$$K_s^*(\text{special}) = 0.805 \pm 0.012 \quad , \phi_s = 0.483 \pm 0.022$$

$$K_b^* = 0.656 \pm 0.004$$

where the errors here and in the rest of the paper are 95% confidence limits. The central value of ϕ_s is an average over the different results for ϕ_s obtained at K_s^*=0.805. In order to obtain the error bars results for K_s^* have been calculated from similar plots based on a smaller number of lines (e.g., N=30-80, 40-80, 50-80). The errors of ϕ_s also take into account the error in the surface temperatures. It is important to note that the values of ϕ for SAWs is significantly larger than the value $\frac{8}{21} \sim 0.38$ obtained by VSS for the θ' model at the special point; therefore our results do not support their conjecture that the θ' and θ points belong to the same universality class.

One expects the partition functions Z_α at (K_b^*, K_s^*) behaves as

$$Z_\alpha = B_\alpha \mu^N N^{\gamma_\alpha - 1}$$

where B_α is a prefactor and μ, the growth parameter, has the same value as for tricritical chains in the bulk. In order to calculate μ and γ_1 we used the same method as for K_s^* and ϕ_s. Thus, at (K_b^*, K_s^*) one obtains (for $\alpha = 1$) $2Z(2N)/Z(N)\mu^N = 2^{\gamma_1}$. Therefore, one can calculate for each pair $(N, 2N)$ the values of $2Z(2N)/Z(N)\mu^N$ for different valus of μ, where the intersection point of these lines should define both γ_1 and the correct μ. A detailed analysis for SAWs leads to

$$\gamma_1(\text{special}) = 1.265 \pm 0.053\,, \qquad \mu(\text{special}) = 3.213 \pm 0.009$$

$$\gamma_1(\text{ordinary}) = 0.598 \pm 0.015\,, \qquad \mu(\text{ordinary}) = 3.212 \pm 0.007$$

Again, the error bars have been obtained by carrying out similar calculations at $(K_b^* \pm D_b, K_s^* \pm D_s)$, where D_b and D_s denote the errors in the corresponding critical temperatures. As expected , the above value of μ is equal, within the error bars, to those obtained for SAWs in the bulk. The values of γ_1 is equal, within the error bars to 1.265 which is significantly larger than the VSS prediction $\frac{8}{7} \sim 1.143$. We also estimated the exponent γ_{11}. As expected, the samples of chains that also end on the surface are relatively small which makes it difficult to use the method described above for γ_1 and μ. Therefore, γ_{11} has been estimated from the values of μ obtained above and best fitting the data for Z_{11} over various ranges of chain lengths $(N_{\text{min}}, N_{\text{max}})$; the smallest value of N_{min} is 20 and the largest values of N_{max} are 60. Thus,

$$\gamma_{11}(\text{special}) = 0.72 \pm 0.06\,, \qquad \gamma_{11}(\text{ordinary}) = -0.46 \pm 0.06\,,$$

The errors have been obtained by carring out such calculations at $(K_b^* \pm D_b, K_s^* \pm D_s, \mu \pm D_\mu)$, where D_μ is the error in μ. The above values of γ_1 and γ_{11} satisfy, within the error bars, the Barber[25] surface scaling relation, $2\gamma_1 - \gamma_{11} = \gamma + \nu$, for the DS values $\frac{8}{7} + \frac{4}{7} = \frac{12}{7} \sim 1.71$.

In summary: This work provides the Monte Carlo estimation of critical exponents of SAWs on the square lattice both at the special and ordinary points. The present results do not support the VSS conjecture that the θ and θ' points belong to the same universality class.[16] We hope that our study will motivate further theoretical work in this still controversial but exciting subject.

We are grateful to Y. Shapir for valuable discussions and for his encouragement. We acknowledge support from the Florida State University Supercomputer Computations Research Institute, which is partially funded by the US Department of Energy under contract number DE-FC05-85ER250000. I. C. acknowledges partial support from the Korea Science and Engineering Foundation through Research Center for Dielectric and Advanced Matter Physics and from Basic Science Research Institute (BSRI-95-2412).

References

[1] P. J. Flory, J. Chem. Phys. **17**, 303 (1949).
[2] P.G. de Gennes, **Scaling Concepts in Polymer Physics**, (Cornell Univ. Press, Ithaca, 1985).
[3] D. Napper, **Polymeric Stabilization of Colloidal Dispersions**, Academic Press N.Y. (1983).
[4] H.S. Chan and K. Dill Annu. Rev. Biophys. Biophys. Chem. **20**, 447 (1991).
[5] B. Nihenhuis, in **Phase Transition and critical Phenomena**, edited by C. Domb and J. Lebowitz (Academic, London, 1987), Vol. 11.
[6] J. L. Cardy, in **Phase Transition and critical Phenomena**, edited by C. Domb and J. Lebowitz (Academic, London, 1987), Vol. 11.
[7] B. Duplantier and H. Saleur, Phys. Rev. Lett. **59**, 539 (1987).
[8] A. Coniglio, N. Jan, I. Majid and H.E. Stanley, Phys. Rev. B **35**, 3617 (1987).
[9] P.H. Poole, A. Coniglio, N. Jan and H.E. Stanley, Phys. Rev. Lett. **60**, 1203 (1988); B. Duplantier and H. Saleur,*ibid*,1204.
[10] P.H. Poole, A. Coniglio, N. Jan and H.E. Stanley, Phys. Rev. **B39**, 495 (1989).
[11] S. O. Warnaar, M.T. Batchelor and B. Nienhuis J. Phys. A **25**, 3077 (1992). This paper proposes the values, $\nu = \frac{12}{23} \sim 0.52$ and $\gamma = \frac{53}{46} \sim 1.15$ for the θ transition. However, these exponents are now thought to describe a transition in a different model; B. Nienhuis, private communication.
[12] F. Seno and A.L. Stella, J. Phys. (France) **49**, 739 (1988); Europhys. Lett., **7**, 605 (1989).
[13] H. Meirovitch and H.A. Lim, J. Chem. Phys., **91**, 2544 (1989).
[14] Vanderzande, Phys. Rev. A **41**, 914 (1990).
[15] A.R. Veal, J.M. Yeomans and G. Jug, J. Phys. A **24**, 827 (1991).
[16] C. Vanderzande, A.L. Stella and F. Seno, Phys. Rev. Lett. **67**,2757 (1991).
[17] H. A. Lim and H. Meirovitch, Phys. Rev. A **39**, 4176 (1989); H. Meirovitch and H.A. Lim *ibid*, **39**, 4186 (1989).
[18] I.S. Chang, H. Meirovitch and Y. Shapir, Phys. Rev. A **41**, 1808 (1990).
[19] H. Meirovitch, J. Chem. Phys. **89**, 2514 (1988).
[20] K. E. Schmidt, Phys. Rev. Lett. **51**, 2175 (1983).
[21] J. M. Hammersley and D. C. Handscomb, **Monte-Carlo Method** (Mathuen, London, 1964).
[22] E. Eisenriegler, K. Kremer and K. Binder, J. Chem. Phys. **77**, 6296 (1982).
[23] S. Livne and Hagai Meirovitch, J. Chem. Phys. **88**, 4498 (1988); Hagai Meirovitch and S. Livne, *ibid*, **88**, 4507 (1988).
[24] M.N. Barber and W. Selke, J. Phys.**A15**, L617 (1982).
[25] M.N. Barber, Phys. Rev. **B8**, 407 (1973); M.N. Barber, A.J. Guttmann, K.M. Middlemiss, G.M. Torrie and S.G. Whittington, J. Phys. A **11**, 1833 (1978).

Oscillating Chemical Reactions Simulated by Molecular Dynamics

K. Geisshirt[1], *E. Praestgaard*[1], *and S. Toxvaerd*[2]

[1]Department of Life Sciences and Chemistry, Roskilde University, DK-4000 Roskilde, Denmark

[2]Department of Theoretical Chemistry, University of Copenhagen, DK-2100 Copenhagen, Denmark

We outline how chemical reactions can be simulated by Molecular Dynamics, and we apply our technique to an oscillating chemical reaction scheme; a modified Volterra-Lotka scheme. We observe deviations from the phenomelogical equations in the phase separating temperature region.

1 Introduction

The phenomenon of oscillating chemical reactions is fairly new, i.e. it has been known for the last three or four decades. The development of the theory for oscillating chemical reactions is strongly connected to the theory of nonlinear dynamical systems and nonequilibrium thermodynamics. Until now oscillating chemical reactions have experimentally been investigated macroscopicly and the theory used for describing the reactions is phenomelogical. Experimentally studied oscillating reactions are either external driven systems or systems allowed to relax to equilibrium.

On the other hand oscillating systems have been investigated numerically as long as oscillating reactions have been known by means of coupled differential equations in the consentrations. It is however possible by today's computers to simulate chemical reactions in continious space and on a molecular level by Monte-Carlo (MC) and Molecular Dynamics (MD) simulations. It implies that it is possible to investigate realistic systems where also other nonlinear dynamics, like a phase seperation might influence the chemical reaction. It was recently shown that chemical reactions have a large effect on phase separation; see e.g. [4] for a Monte-Carlo study of simple first-order reactions $A \rightleftarrows B$ and Molecular Dynamics study of first- and second-order reactions [5].

2 Chemical kinetics and molecular dynamics

Chemical reactions, on a microscopic level, are usually formulated as coupled elementary reactions. We will here only consider coupled bimolecular reactions, i.e. a reactions where there are two reactants in each reactions.

The usual chemical viewpoint is the transition state theory. The basic idea is that the two reactants collide and form an activated complex ($C^{\ddagger}$), which can either break into the reactants again or into the products. The quantum mechanical details in this reversible dynamics are, however, believed to be of no

Springer Proceedings in Physics, Volume 82
Computer Simulation Studies in Condensed-Matter Physics IX
Eds.: D. P. Landau, K.K. Mon, H. -B. Schüttler

importance for the overall kinetics of oscillating reactions. The main feature is that the exchanges of covalent bonds occur with a certain probability when the reactants are separated at a short molecular distance. The reaction rate is typically given by an Arrhenius expression where the rate depends exponentially on the activation energy in units of the kinetic energy (the temperature). Experimentally, reactions with small activation energy are difficult to follow because they are fast. In Molecular Dynamics, however, we are in the opposite situation. With today's computers we are able to simulate of the order 10^5 (simple) molecules in of the order 10^6 time steps, which correspond to reaction times of only nanoseconds. For this reason the transition energy is taken to be of the order T which automatically implies that the reaction takes place on a time scale equal to the mean collision time. Futhermore, a chemical reaction appears in an open system, i.e. without a conserved energy and fixed number of reactant. On the other hand a traditional MD system consists of a fixed number of particles and with a well-defined Hamiltonian, which makes it to a non-trivial task to reformulate the dynamics in open systems with chemical reactions. The actual implementation and its impact on the dynamics is described in in more details in Ref. [5]. In summary the reaction dynamics is implemented into the MD by spontaneously and with a certain probability, to relabel the colliding particles. Futhermore, in order not to introduce large force gradients into the system by this reaction dynamics we only consider reactions between particles with almost the same excluded volumes, i.e. the chemical individualities of the reactants are given by their long range forces and not by their short range forces. This feature is certainly unrealistic from a chemical view point; but it is believed to be of no importance for the qualitative behavior of the chemical reaction dynamics as will be demonstrated.

3 A particular example: MD simulations of the modified Volterra-Lotka reaction

We have chosen a particular oscillating system as our system. It consists of three species and three reactions. The reactions are:

$$A + X \xrightarrow{k_1} 2X \tag{1}$$

$$X + Y \xrightarrow{k_2} 2Y \tag{2}$$

$$Y + A \xrightarrow{k_3} 2A \tag{3}$$

To minimise the number of parameters in our model, we have chosen to set k_3 to the same value as k_1.

This reaction scheme is a simple modification of the Volterra-Lotka equation for a homogeneous (in space) oscillating reaction [6] [7]. In a traditional chemical reaction experiment the system is fed by a reactant A which through a consecutive set of unimolecular reactions is transformed into the product B. In our case and for an open MD system of particles in a volume with periodical boundaries it is, however, convenient to equal reactant and product and thus to recover A through reaction 3. It is the only modification of the traditional reaction scheme for the oscillating reaction and the flow through the system is obtained, as usual by only forward elementary reactions without reverse reactions. This corresponds to that the reactant has a significant higher (free) energy than the product.

From these three reactions, one can easily derive three phenomelogical equations describing the concentrations of the three species. The equations are:

$$\begin{aligned}\frac{d[A]}{dt} &= -k_1[A][X] + k_1[A][Y] \\ \frac{d[X]}{dt} &= k_1[A][X] - k_2[X][Y] \\ \frac{d[Y]}{dt} &= k_2[X][Y] - k_1[A][Y]\end{aligned}$$

where $[\cdot]$ means the concentration. The phenomelogical equations are only valid in the complete homogeneous case, which experimentally is realised by stirring the contents of the chemical reactor.

We have three coupled ordinary differential equations which are impossible to find an analytical solution to. In general, we have equations of the type:

$$\frac{d\vec{x}}{dt} = \vec{f}(\vec{x})$$

We can first find a stationary solution to the problem above, i.e. find an $\vec{x}_0$ which satisfies $\vec{f}(\vec{x}_0) = \vec{0}$. Then we can linearise the differential equation at this stationary point and obtain

$$\frac{d\delta\vec{x}}{dt} = \mathbf{J}(\vec{x}_0)\delta\vec{x}$$

where $\mathbf{J}$ is the Jacobian matrix and $\delta\vec{x}$ is the deviation from the stationary point, i.e. $\delta\vec{x} = \vec{x} - \vec{x}_0$. The linearised system is a linear differential equation. The solution of the linear differential equation is

$$\delta\vec{x} = \sum_i \vec{e}_i e^{\omega_i t}$$

where ω_i is the ith eigenvalue and $\vec{e}_i$ is basically the ith eigenvector which is normalised appropriately.

The modified Volterra-Lotka scheme has a stationary solution, which is

$$\begin{aligned}[A]_0 &= C\left(1 - \frac{2k_1}{2k_1 + k_2}\right) \\ [X]_0 &= [Y]_0 = \frac{Ck_1}{2k_1 + k_2}\end{aligned}$$

where C is the total concentration, i.e. $C = [A] + [X] + [Y]$ which is a constant.

We can now linearise the system and calculate the eigenvalues of the Jacobian matrix at the stationary point. We get

$$\lambda_1 = \overline{\lambda_2} = \frac{Ck_1\sqrt{k_2}}{\sqrt{2k_1 + k_2}}\imath$$

which shows us that the system will have sustained oscillations - at least close to the steady-state solution.

4 Simulational details

The chemical equations for the Volterra-Lotka reaction expresses a *homogeneous* change in the concentration space of the reactants, which for certain values of the reaction rates, oscillate. In a real experiment it means that large concentrations of a certain component is built up for a while and then disappears through a chemical reaction. If the three components are miscible for all concentrations one would expect that the dynamics might be well described by the traditional homogeneous differential equations, whereas a new situation appears if one of the concentrations exceeds the critical value of solubility and a competing phase separation takes place in the oscillating system. For this reason we will expect a non-trivial difference between a set of homogeneous chemical reactions and a corresponding MD simulation. This is due to the fact that we in an MD system can create a phase separation by a chemical reaction, simply by ensuring that the product has an intermolecular potential which favors phase separation. This is ensured in the very same manner as for the chemical reaction by varying the long range attractions between the molecules. For details see e.g. [8].

The system is - as already mentioned - a three-component system. The interaction of particles of the same type is a Lennard-Jones potential which is cut at $r = 2.5$. The interaction between particles of different types is also a Lennard-Jones potential, but it is cut at $r = 1.0$ meaning that it is replusive only. The density of the system is 0.6. At low enough temperature a system with such interactions will phase separate. We use the same reaction parameters throughout all our simulations. The collision diameter is set to 1.0 (in Lennard-Jones units). The probability of reaction is set to $1.0 \cdot 10^{-3}$ for reactions 1 and 3 while for reaction 2 it is set to $1.1 \cdot 10^{-3}$.

All our simulations are performed in the NVT-ensemble using a Nosé-Hoover thermostat to control the temperature and a leap-frog algorithm to integrate the equations of motion [1]. The chosen time step is $h = 5 \cdot 10^{-4}$. Furthermore, all simulations are done in two spatial dimensions.

5 Results and open questions

We have performed a number of simulations at different temperatures with either 1024 or 8192 particles. Rate constant for a reaction usually follow Arrhenius' law, i.e. $k(T) = A \cdot e^{-E_a/RT}$, where E_a is the activation energy, A is a preexponetial factor, and R is the gas constant.

During our simulations we have stored instantaneous values of the concentrations and the number of times the three reactions has occured. Figure 1

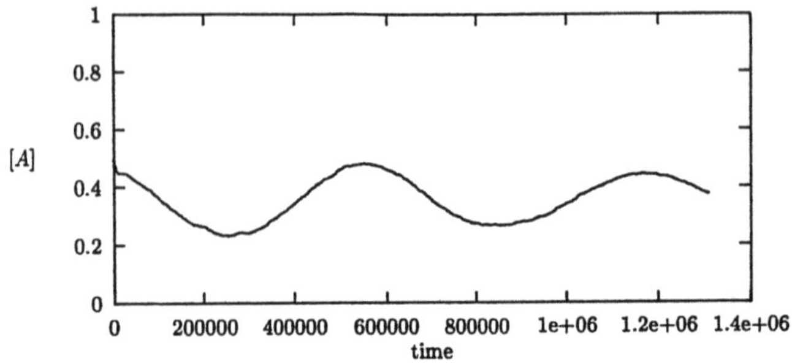

Figure 1: The concentration of A versus time. The simulation shown consists of 8192 particles and the temperature is 1.4. The time is measured in unit of time steps.

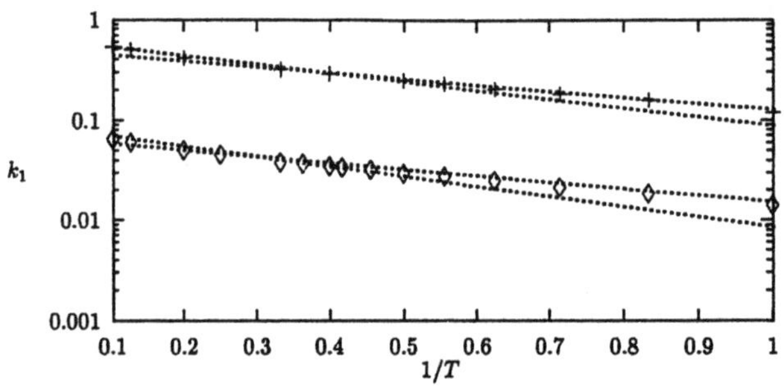

Figure 2: Arrhenius plot for the rate constant for reaction 1. The lower curve is for 1024 particles and the upper one is for 8192 particles. Note that the rate constants are not normalised according to the number of particles and therefore the rate constants for the 8192 particle simulations are 8 times larger. The dotted lines are the best fitted lines to the data in that particular temperature region.

shows the concentration of A versus time for a typical simulation. The number of times a reaction has occured (denoted N_R) is proprotional to the velocity of that reaction and we therefore have a simple relationship for the rate constant, namely $k = \frac{N_R}{[J][J']}$, where J and J' denote the two reactants in the reaction.

Figure 2 shows the rate constant for reaction 1 versus the inverse temperature. Similar plots can be made for the two other reactions. The estimated error bars are very small. Our estimates are that they are at 4th or 5th significant digit [2].

We notice there is a crossover at $T \approx 3$. The crossover seems not just to be a finite-size effect since we see it at simulations with 1024 and 8192 particles.

We believe that this effect comes from the fact that the system goes from being phase separated to a homogeneous phase when we increase the temperature. In the homogeneous phase the three reactions can occur at all places in the simulation box while below the phase separating temperature, the reactions will mainly occur at the boundaries between clusters of, say, X and Y. The critical temperature for a similar system of a binary mixture is $T_c \approx 4.7$ [8]. In the critical region the phase separation dynamics is extremely slow; but well below this region two processes will compete: diffusion (to the interfacial reaction zone) and reaction. The figure shows that the activation energy in the low temperature region is not the activation energy for the reaction but the activation energy for combined process of diffusion and reaction.

Acknowledgements

The authors wish to thank the National Science Foundation (in Denmark) for computing time on the IBM SP2 at the National Computer Centre (Uni-C). We also wish to thank M. Laradji for his many advices.

References

[1] Toxvaerd, S., 1991, Mol. Phys. **72**(1), 159.

[2] Flyvbjerg, H. and Petersen, H.G., 1989, J. Chem. Phys. **91**(1), 461.

[3] Chemical Kinetics, 3rd edition. K.J. Laidler. HarperCollins, 1987.

[4] Glotzer, S.C., Stauffer, D., and Jan, N., 1994, Phys. Rev. Lett. **72**(26), 4109.

[5] Toxvaerd, S., 1996, *to appear in* Phys. Rev. E **53**.

[6] Lotka, A., 1910, J. Phys. Chem. **14**, 271.

[7] Lotka, A., 1920, J. Am. Chem. Phys. **42**, 1593.

[8] Toxvaerd, S. and Velasco, E., 1995, Mol. Phys. **86**, 845.

Spinodal Decomposition in Three-Dimensional Binary Fluids: A Large-Scale Molecular Dynamics Simulation

M. Laradji[1], *S. Toxvaerd*[2], *and O.G. Mouritsen*[3]

[1]Department of Physics, University of Toronto, Toronto, Ontario, Canada M5S 1A7

[2]Department of Chemistry, H.C. Ørsted Institute, University of Copenhagen, DK-2100 Copenhagen, Denmark

[3]Department of Physical Chemistry, The Technical University of Denmark, Building 206, DK-2800 Lyngby, Denmark

Abstract. The dynamics of phase separation in three-dimensional binary fluids is investigated by means of extensive molecular dynamics simulations. We found a dynamical scaling behaviour at late times with a growth exponent $n = 1$, in agreement with the dimensional analysis for viscous growth.

1. Introduction

The dynamics of domain coarsening in binary fluids following a temperature quench from the high-temperature homogeneous phase to the coexistence region of the phase diagram has attracted a lot of attention during the past few years. As a consequence of the nonlinearity of the dynamical equations relevant for these systems, most of the theoretical studies were computer investigations based on numerical integration of the relevant time-dependent Ginzburg-Landau model (model H) [1-4], lattice-Boltzmann (LB) simulations [5] and molecular dynamics (MD) simulations [6]. Unlike binary alloys, where the late-time domain growth is mainly due to the long-range evaporation-condensation of material, and is characterized by a power-law time-dependence of the average domain size, $R(t) \sim t^n$, with an exponent $n = 1/3$ [7], the dynamics in binary fluids is more complicated due to the coupling of the velocity field (hydrodynamic modes) to the ordering field [8,9]. Dimensional analysis of model H leads to a diffusive growth at intermediate times, $R(t) \sim (\gamma t)^{1/3}$, followed by a viscous regime at later times, $R(t) \sim \gamma t/\eta$, and at very late times, this regime is followed by an inertial growth, $R(t) \sim (\gamma t^2/\rho)^{1/3}$, where γ, η, ρ are the surface tension, the shear viscosity and the fluid density respectively. The crossover from the viscous regime to the inertial regime occurs at a length scale, $R(t) \sim \eta^2/\gamma\rho$. The viscous regime was predicted, for the first time, by Siggia [8], and is due to a necking-down instability of the interconnected structure. On the other hand, the physical process lading to the $t^{2/3}$-growth is not clear. Several experiments

Springer Proceedings in Physics, Volume 82
Computer Simulation Studies in Condensed-Matter Physics IX
Eds.: D. P. Landau, K.K. Mon, H. -B. Schüttler

on simple binary fluids and on homopolymer binary mixtures find a viscous growth during the late stages of the phase separation, $n = 1$, but an inertial growth has never been observed [10,11]. Numerically, both model H and LB simulations seem to confirm the experimental picture in three dimensions [1–5]. However, a recent MD simulation by Ma *et al.* showed that the asymptotic growth regime is rather inertial with a growth exponent $n = 2/3$ [6]. In order to elucidate this apparent discrepancy between the MD simulation of the binary Lennard–Jones potential and the other simulations and experiments, we have conducted a systematic and extensive parallel MD simulations on large systems and during periods of time which are much longer than in the recent MD simulations [6]. Our systematic study shows that the growth is viscous, in contradiction to the simulation of Ma *et al.*. We also observed a clear dynamical scaling prevailing at late times.

2. Model and Detail of the Numerical Simulation

In our simulation model, we consider N monoatomic molecules belonging to two species. Molecules belonging to the same species, interact via the standard 12–6 Lennard–Jones (LJ) potential, truncated and shifted at a cutoff distance $r_c = 2.5\sigma$. Molecules belonging to different species also interact via the same potential but truncated and shifted at $r_c = 2^{1/6}\sigma$. This choice of the potential ensures that the interaction between particles of different species is only repulsive. The Hamilton equations are integrated using the Leap–frog algorithm with a time step of $\Delta t = 0.005\tau$ where the time scale is $\tau = \sqrt{\mu\sigma^2/\epsilon}$, and the temperature is controlled by a Nosè–Hoover thermostat. We have considered a large number of particles, $N = 343,000$, with a fluid density, $\rho = 0.8\sigma^{-3}$. The choice of this density ensures that we are safely within the single fluid phase. The binary fluid at this density exhibits a coexistence region with a critical point at $k_BT_c = (4.7 \pm 0.2)\epsilon$ [12]. We have performed critical quenches at the following temperatures $k_BT = 1$, 2, 3, 3.5, 3.75 and 4 ϵ, and we have also performed other off critical quenches. For each system, many independent runs were performed (14 for $k_BT = 2\epsilon$ and 4 for all others) [13]. Our simulations were performed on a SP2 parallel computer using 12 processors [14].

We have calculated the pair correlation function $G(\mathbf{r}, t) = \langle\phi(\mathbf{r}, t)\ \phi(0, t)\rangle$, where the order parameter $\phi(\mathbf{r}, t)$ is the difference between the local volume fractions of the two components, and the structure factor, $S(\mathbf{q}, t) = \langle|\phi(\mathbf{q}, t)|^2\rangle/V$. The average domain size is calculated from two definitions: first, as the inverse of the first moment of the structure factor, $R_1(t)$, and second as the first zero of the correlation function, $R_2(t)$.

3. Domain Growth

In Fig. 1(a), the average domain size is plotted as a function of time for a quench at $k_BT = 2\epsilon$. This data shows very clearly that at late times the growth is linear with time. However, a log–log plot of our data seems to show a growth law with an asymptotic exponent $n = 2/3$, as shown in Fig. 1(b) for all temperatures considered by our simulations. However, we should notice that a growth law, $R(t) = R(0) + at$, would exhibit an effective growth exponent which could be much smaller than 1 due to the large value of $R(0)$ and may display other non–algebraic dependences in time.

In order to determine the true asymptotic growth exponent, we calculated numerically the values of the surface tension and the shear viscosity, and plotted $R(t)$ versus $\tilde{t}_v = (\gamma t/\eta)$ in Fig. 2(a), and versus $\tilde{t}_i = (\gamma/\rho t^2)^{1/3}$ in Fig. 2(b). Clearly, the growth cannot be inertial, since the prefactor of $\tilde{t}_i$ depends on temperature. However, Fig. 2(a) shows very clearly that all curves have the same asymptotic slope as a function of $\tilde{t}_v$ in agreement with the scaling arguments of viscous growth. We therefore conclude that the asymptotic growth exponent, in the critically quenched binary Lennard–Jones potential, is $n = 1$. This is in contradiction to the MD simulation of Ma *et al.* [6], but agrees with both previous LB and model H simulations as well as with experiments. Moreover, the scaling arguments show that the prefactor does not depend on the volume fraction, as long as the domain structure is interconnected. Indeed the off–critical quenches performed by us showed the same slope

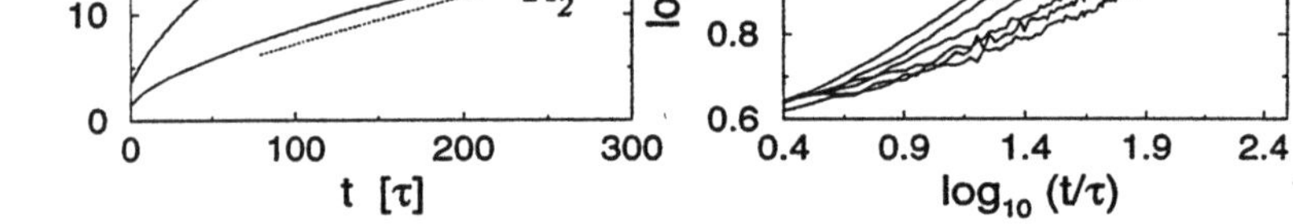

Figure 1: (a) Average domain size as a function of time for $k_BT = 2\epsilon$. The upper solid line is calculated from the first moment of $S(q,t)$, and the lower one is calculated from the first zero of $G(r,t)$. The two dotted lines are straight lines; (b) log$R(t)$ versus log$t$. Solid curves from top to bottom correspond to $k_BT = 1$, 2, 3, 3.5, 3.75 and 4ϵ respectively. The dotted line has a slope of 2/3

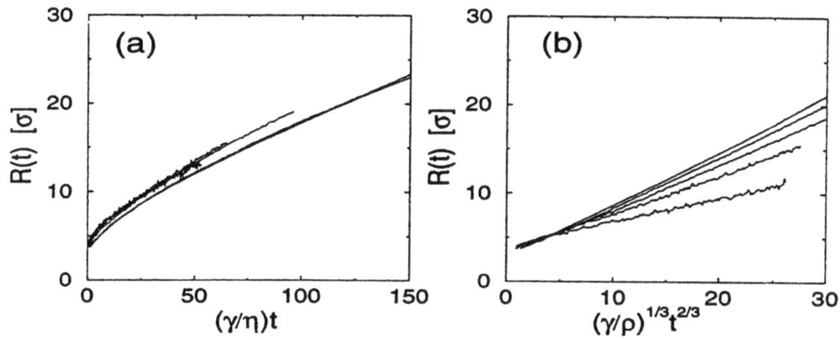

Figure 2: (a) The average domain size as a function of $\tilde{t}_v$ for all systems investigated. (b) The average domain size as a function of $\tilde{t}_i$. Curves from top to bottom correspond to k_BT = 1, 2, 3, 3.5, 3.75 and 4ϵ respectively.

as a function of $\tilde{t}_v$ for volume fractions up to 0.15. The prefactor of the $(\gamma/\eta)t$ in the inertial growth was esimated by Siggia to be roughly 0.6 [8]. On the other hand, the linear stability analysis by SanMiguel *et al* [15] predicts a prefactor of 0.25. We estimated this coefficient from our data to be 0.11 ± 0.01. This value is quite different from the theoretical predictions, which we should note are quite crude. However, our value compares very well with the experimental value of Guenoun *et al.*, 0.138 ± 0.006 [10].

4. Dynamical Scaling

During the late times of the phase separation, the average domain size becomes the only dominant length scale in the system. Consequently, a rescaling of all lengths with the average domain size, leads to structures which, on the average, look similar at different times. Quantitatively, this translates into a time–independent pair–correlation function, $g(\chi) = G(r,t)$, where $\chi = x/R(t)$ is the scaling length, or into a scaling structure factor $F(x) = S(q,t)/R(t)^3$ where $x = qR(t)$ is the scaled wave–vector. In the inset of Fig. 3(a), the time–evolution of the correlation function is shown. The increase of the value of the first zero of the correlation function indicates that the domains are coarsening. The scaled correlation function is shown in Fig. 3(a). Indeed, we observe that starting from $t = 80\tau$ all data collapse into a universal time–independent correlation function. Very nice scaling is also observed in the structure factor as shown in Fig. 3(b).

At late times, the interfacial width becomes very small compared to the average domain size, this leads to a power law depen-

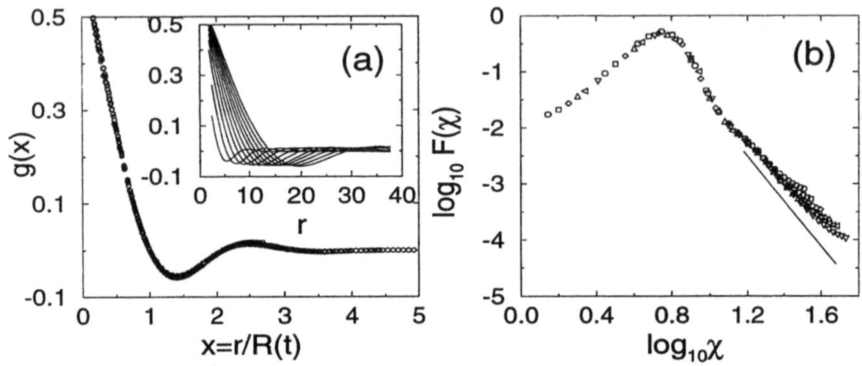

Figure 3: (a) The pair–correlation function versus the scaled length, $r/R(t)$ fo $k_BT = 2\epsilon$. Data shown are from $t = 80$ to 220τ. The time evolution of the pair–correlation function is shown in the inset of (a). Data from left to right is from $t = 20$ to 220τ in steps of 20τ. In (b) the scaling structure factor is shown for the same system. The straight line has a slope of -4.

dence of the structure factor with x, $F(x) = x^{-(d+1)}$. This is known as Porod's law. Indeed, Fig. 3(b), shows very clearly that our data complies nicely with Porod's law.

5. Conclusion

Spinodal decomposition in three–dimensional binary fluids is investigated by means of extensive and systematic molecular dynamics simulations. We found that during late times, the domain growth is viscous with a growth exponent $n = 1$, in agreement with previous Langevin and lattice–Boltzmann simulations as well as experiments. However, our results contradict the recent molecular dynamics simulations of Ma *et al.* [6] who observed an inertial regime with an exponent of $n = 2/3$. We have also observed a dynamical scaling behaviour during the late times.

Acknowledgements

The authors would like to thank R.C. Desai for useful discussions, and P. Padilla for her help in calculating the surface tension and the shear viscosity. This work is supported by the Natural Sciences and Engineering Research Council of Canada, by the Danish Natural Science Research Council and by the Danish Technical Research Council.

References

[1] T. Koga and K. Kawasaki, Phys. Rev. A **44**, R817 (1991).
[2] S. Puri and B. Dünweg, Phys. Rev. A **45**, R6977 (1992).
[3] O.T. Valls and J.E. Farrell, Phys. Rev. E **47**, R36 (1993).
[4] A. Shinozaki and Y. Oono, Phys. Rev. E **48**, 2622 (1993).
[5] F.J. Alexander, S. Chen, and D.W. Grunau, Phys. Rev. B **48**, 634 (1993).
[6] W.-J. Ma, A. Maritan, J.R. Banavar, and J. Koplik, Phys. Rev. A **45**, R5347 (1992).
[7] I.L. Lifshitz and V.V. Slyozov, J. Phys. Chem. Solids **19**, 35 (1962).
[8] E.D. Siggia, Phys. Rev. A **20**, 595 (1979).
[9] H. Furukawa, Adv. Phys. **34**, 703 (1985).
[10] P. Guenoun, R. Gastaud, F. Perrot, and D. Beysens, Phys. Rev. A **36**, 4876 (1987).
[11] N. Kuwahara, H. Sato, and K. Kubota, Phys. Rev. E **47**, 1132 (1993).
[12] S. Toxvaerd and K. Velasco, Mol. Phys. **86**, 845 (1995).
[13] More detail of our simulation and other results will be published elsewhere.
[14] The simulations were performed at UNI–C, the Danish supercomputing centre.
[15] M. San Miguel, M. Grant, and J.D. Gunton, Phys. Rev. A **31**, 1001 (1985).

Simulation of a Reacting Plasma: Preliminary Results

B. Foley[1], *A.M. Ferrenberg*[2], *and N.G. Adams*[1]

[1]Department of Chemistry, The University of Georgia, Athens, GA 30602, USA

[2]University Computing and Networking Services and Center for Simulational Physics, The University of Georgia, Athens, GA 30602, USA

Abstract. Electron-ion recombination is important to the study of any process involving plasmas. It provides a primary ionization-loss mechanism and a source of many highly excited and radical species. Using current techniques, detailed analysis of light emissions from this process is difficult. We present preliminary results of simulations of reacting plasmas which, eventually, should allow for a more quantitative analysis of experimental results.

I. Introduction

Our work focuses on the study of emissions resulting from the electron recombination of small, polyatomic ions. Electron-ion recombination is the neutralization of a positive ion, here, CO_2^+, by an electron (e^-). The process ($CO_2^+ + e^- \rightarrow CO^* + O$) usually breaks the neutralized ion into fragments which are often in electronically excited states ($CO^* \rightarrow CO + h\nu$).

We can determine much about the nature of this process by studying the light emitted from these fragments. Emissions from electron recombination have been studied previously [1, 2, 3, 4, 5], however, relative concentrations of the various electronically excited states produced in these reactions have not been determined to more than one digit of certainty. It is hoped that a simulation of the reaction region will facilitate determination of these relative populations.

II. Brief Experimental Technique

Since the primary focus of these proceedings is computational, the experimental technique will only be discussed briefly. More detailed descriptions of this and similar techniques can be found elsewhere [6, 7].
The apparatus and technique are summarized in Figure 1, and pertinent parameters are given in Table 1.

Through the series of reactions summarized in [6, 7] and in Fig. 1, the flow is converted into a mixture primarily composed of ionized argon and electrons in a neutral helium carrier gas. CO_2 is then injected into this flow. Ar^+ charge-transfers to the CO_2 non-destructively, meaning that one non-

Springer Proceedings in Physics, Volume 82
Computer Simulation Studies in Condensed-Matter Physics IX
Eds.: D. P. Landau, K.K. Mon, H. -B. Schüttler

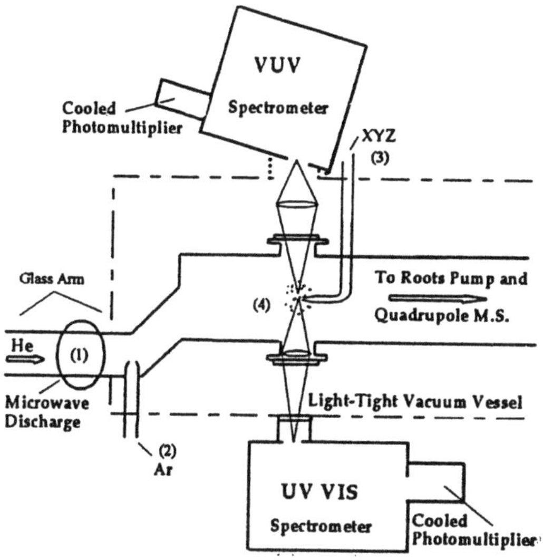

Fig. 1: (1) He ionized by μ-wave discharge source. (2) Ar converts flow to Ar^+/e plasma. Addition of CO_2 (3) yields CO_2^+/e plasma that recombines (4).

Table 1

Approximate Densities	
Neutral He:	6×10^{16} cm^{-3}
Neutral Ar:	1×10^{13} cm^{-3}
Neutral CO_2:	*varies with position*
Ionized Ar:	1×10^{10} cm^{-3}
Upstream Ionization:	$\sim 4 \times 10^{10}$ cm^{-3}
Downstream Ionization:	$\sim 1 \times 10^{10}$ cm^{-3}
All other consituents:	$< 1 \times 10^{10}$ cm^{-3}
Flowtube Construction and Characteristics	
Flowtube diameter:	~ 10 cm
Length:	1 meter
View window diameter:	2.54 cm
Spectrometer range:	100–800 nm
Pressure:	1.8 Torr
Temperature:	298 K

recombining ion (Ar^+) is exchanged for one recombining ion (CO_2^+). Since this happens mid-flow, the recombining ion is produced in the midst of "cold," ambient electrons. The entire process occurs very rapidly, so any emissions produced within the appropriate region are collected by two spectrometers for vacuum ultraviolet (VUV) and ultraviolet/visible (UV/VIS) detection. (see Table 1).

III. Typical Experimental Results and Simulation Rationale

For a spectrum containing any set of species having the same radiative lifetimes (for example, vibrational states within one electronic state), determination of the relative populations is possible with little further analysis (see refs [6, 7, 8, 9] for examples of this sort of population analysis). in However, to precisely determine contributions from species with varying radiative lifetimes, it is necessary to know exactly where and when the species emitted its light.

Photographs (black and white reproductions from original color) of the reaction region for three flows of CO_2 injected against a constant counter–flow of He/Ar^+/e are provided in Figure 2, and two spectra taken under conditions similar to (a) and (c) are given in Figure 3. Figure 2 shows the progression of the bright 'reaction front' across the viewport as the CO_2 throughput is increased, eventually moving it out of the viewing region altogether. In a color print of the middle flow (b), The visible portion of the 'front' is comprised mainly of the 'white' emissions, with reddish-brown emissions trailing off downstream. In low-flow photos (a), hardly any red-brown can be seen, and in high-flow photos (c), the red-brown is very prominent. This is consistent with the CO_2^+ radiating quickly ($\tau \sim 140$ ns) at the front, and the CO d $^3\Delta_i$ and CO^+ $^2\Pi_i$ ($\tau \sim 7500$ ns and 3800 ns, respectively) states radiating as they travel downstream.

(a) (b) (c)

Fig. 2: Photographs of a recombining CO_2^+/e plasma. Note effect of increasing CO_2 input (a<b<c, out of port, from right to left in these photos) against a constant counterflow of He (from left to right). Compare (a) and (c) to Fig. 3. Time in viewing region $\sim$ 1 ms.

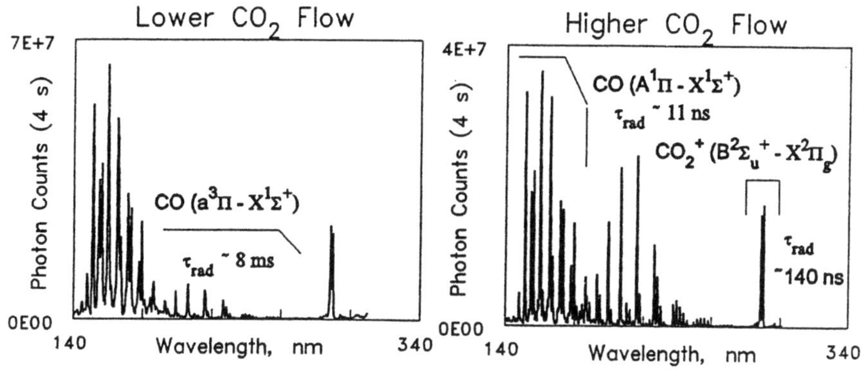

Fig. 3: Effect on observed spectrum of changing CO_2 input. Compare to photos (a) and (c) of Fig. 2.

The emissions represented in Fig. 3 are not visible, but the effects of changing the flow are still evident. At low flow, the spectrum in Fig. 3 is dominated by 'faster' emissions from CO_2^+ and the First Positive (A–X) system of CO. At higher flows, however, the 'slower' Cameron Bands (a–X) play a much more significant role. More drastic differences in spectral intensity can be generated by manipulating the flow conditions.

Since the distributions of emitting species in the flow are not isotropic with respect to the viewing region of the spectrometer, interpretation of the spectra is not trivial. Detailed descriptions of the species density distributions in the reaction region are necessary. Experimental determinations are prohibitively costly for our current circumstance. Density distributions determined via simulation will be coupled with reaction rates and radiative lifetimes to mimic the visible appearance of the plume, and provide information regarding the distributions of the invisible components of the reacting plume.

IV. Simulation Method and Preliminary Results

For the preliminary phase of this project, we have developed an event-driven molecular dynamics[10] program to study the gas flow at the microscopic level. Considering the relatively low densities of the gases involved, we hope that such an approach can provide meaningful results with reasonable computing effort. If this approach should prove too challenging computationally, the information it provides will, in any case, prove invaluable in the development of macroscopic solutions to the problem, either by hydrodynamic methods[11] or via the Boltzmann equation[12].

We model both the He atoms as well as the CO_2 molecules as hard spheres. For He, this is a very good approximation; however, for CO_2, a linear molecule, the approximation is less effective. To compensate for the non-spherical

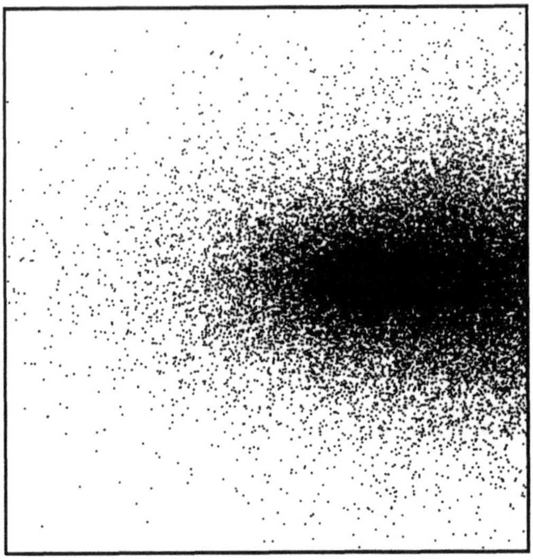

Fig. 4: "Snapshot" of CO_2 molecule positions, projected onto the x-y plane, as generated by our simulation method. Orientation has been chosen such that the appearance mimics the photos in Fig. 2.

shape of the CO_2, we used an effective radius calculated from the molecular volume of the molecule, then randomized the direction of collisions in the center of mass coordinate system. The radii for the molecules are 0.979 Å and 1.16 Å for He and CO_2 respectively. The CO_2 gas is injected in the system with a velocity distribution that consists of the Maxwell distribution at 300 K, shifted in the z-direction by a uniform flow velocity, then truncated to consist of only positive z-velocities. (CO_2 molecules with negative z-velocity are unable to exit the port).
He molecules are introduced with a similar velocity distribution (but with a flow velocity in the negative z-direction) within the top 10 % of the box. Snapshots of the CO_2 molecule positions, projected into the xz-plane are shown in Figure 4.

In the early stages of this project, we kept track of all collisions occurring in the system. However, because the average He density is so much larger than the CO_2 density, this requires a very large number of particles in the simulation to have sufficient CO_2 molecules for analysis. To overcome this difficulty, we replaced the real He molecules with a field of "virtual" molecules which are created when a collision between He and CO_2 would take place, and disappear after the collision. The collision rate for a CO_2 molecule in the presence of many He molecules (all represented by hard spheres) can be expressed simply as $\pi \cdot (R_{CO_2} + R_{He})^2 \cdot \rho_{He} \cdot V$ where ρ_{He} is the He density and V is the average relative speed. We calculate the average relative speed using the actual velocity of the CO_2 molecule (V_x, V_y, V_z) and the known distribution of He velocities

$$V = (< V_{He}{}^2 > + V_x^2 + V_y^2 + V_z^2 - 2 \cdot V_z \cdot V_{flow})^{1/2}$$

This approach increased the speed of the program by about three orders of magnitude (since it eliminated the He molecules which have a mean density approximately 1000 times greater than the CO_2 density). Results obtained with this program agree very well with those obtained with the original program. As a final approximation, we ignored CO_2-CO_2 interactions which occur only rarely. With the resulting program, we are able to easily study systems with 50000 to 100000 CO_2 molecules. It will be difficult, however, to determine exactly how turning off the CO_2-CO_2 collisions affects the results.

V. Conclusions

Our preliminary results are encouraging, but much work remains to be done. We believe that these simulations can lead to realistic estimates of static properties of the flow (such as the CO_2 density and velocity distributions). Incorporating the dynamic processes involved in the actual reactions will prove to be more challenging.

Acknowledgments

We wish to thank Burkhard Dünweg for stimulating discussions. Financial support of the NSF (Division of Astronomical Sciences) under Grant AST-9415485 is gratefully acknowledged.

References

1. T. S. Wauchop and H. P. Broida, *J. Chem. Phys.* **56(1)**, 330 (1972).
2. F. Vallee, B. R. Rowe, J. C. Gomet, J. L. Queffelec and M. Morlais, *Chem. Phys. Lett.* **124(4)**, 317 (1986).
3. R. A. Gutcheck and E. C. Zipf, *J. Geophys. Res.* **78(25)** 5429 (1973).
4. C. A. Weller and M. A. Biondi, *Phys. Rev. Lett.* **19(2)**, 59 (1967).
5. M. Tsuji, M. Nakamura, Y. Nishimura and H. Obase, *J. Chem. Phys.* **103(4)**, 1413 (1995).
6. B. L. Foley, N. G. Adams and H. S. Lee, *J. Phys. Chem.* **97**, 5218 (1993).
7. N. G. Adams, *Int. J. Mass Spec. Ion Proc.* **132(1-2)**, 1 (1994).
8. N. G. Adams and L. M. Babcock, *Ap. J.* **434**, 184 (1994).
9. N. G. Adams and L. M. Babcock, *J. Phys. Chem.* **98**, 4564 (1994).
10. D. C. Rapaport, *The Art of Molecular Dynamics Simulation.* Cambridge University Press, New York. (1995).
11. A. Uehigashi, S. Sugiura, K. Morinishi and N. Satofuka, *JSME Int. J. Ser. II.* **35(4)**, 507 (1992).
12. L. Boltzmann (S. G. Brush, tr), *Lectures on Gas Theory.* University of California Press, Berkeley. (1964).

Behaviour of the Finite-Sized, Three-Dimensional, Ising Model Near the Critical Point

G.A. Baker Jr. and R. Gupta

Theoretical Division, Los Alamos National Laboratory,
University of California, Los Alamos, NM 87545, USA

Abstract Recent work showing the validity of hyperscaling involved results for finite size systems very near the critical point. We study this problem in more detail, and give estimators related to the Binder cumulant ratio which seem to approach the critical temperature from above and below. Based on these results, we estimate that the renormalized coupling constant, computed for the temperature fixed at the critical temperature and then taking the large system-size limit, is about 4.9 ± 0.1, and give a likely lower bound for it of 4.5. These estimates are argued to suffice to show the validity of hyperscaling.

The standard, sample problem in the study of critical phenomena is the Ising model. At a critical point in, for example, the temperature – magnetic field plane, many of the thermodynamic properties of a magnet become singular like some, often non-integer, power of a measure of the distance to the critical point in this plane. There are a number of these properties and the exponents describing their singular behavior are called critical indices. One of the long-standing problems in the study of the Ising model has been the validity of hyper-scaling. This property relates to the question of whether the relations between the critical indices which involve the spatial dimension hold or not. One of the key measures of the validity of hyperscaling is whether the quantity,

$$g(K) \equiv - \lim_{L\to\infty} \left(\frac{v}{a^d}\right) \frac{\frac{\partial^2 \chi(K)}{\partial H^2}}{\chi^2(K)\xi^d(K)}, \tag{1}$$

vanishes as the inverse temperature K approaches the critical-point value K_c for the magnetic field H at its critical value of zero. Here d is the spatial dimension, v is the volume of a lattice cell, a is the lattice spacing, L is the system size, χ is the magnetic susceptibility, and ξ is correlation length. If $\lim_{K\to K_c} g(K) \equiv g^* = 0$, then hyperscaling may fail, otherwise it holds.

In his study by exact calculation of the two-dimensional, Ising model, Baker [1] found that the critical point in the temperature – system-size plane is a point of non-uniform approach, *i.e.*,

$$g^* = \lim_{K\to K_c} \lim_{L\to\infty} g(K,L) \neq \lim_{L\to\infty} \lim_{K\to K_c} g(K,L) = \hat{g}. \tag{2}$$

The first of these two limits is the thermodynamically relevant one. Baker and

Springer Proceedings in Physics, Volume 82
Computer Simulation Studies in Condensed-Matter Physics IX
Eds.: D. P. Landau, K.K. Mon, H. -B. Schüttler

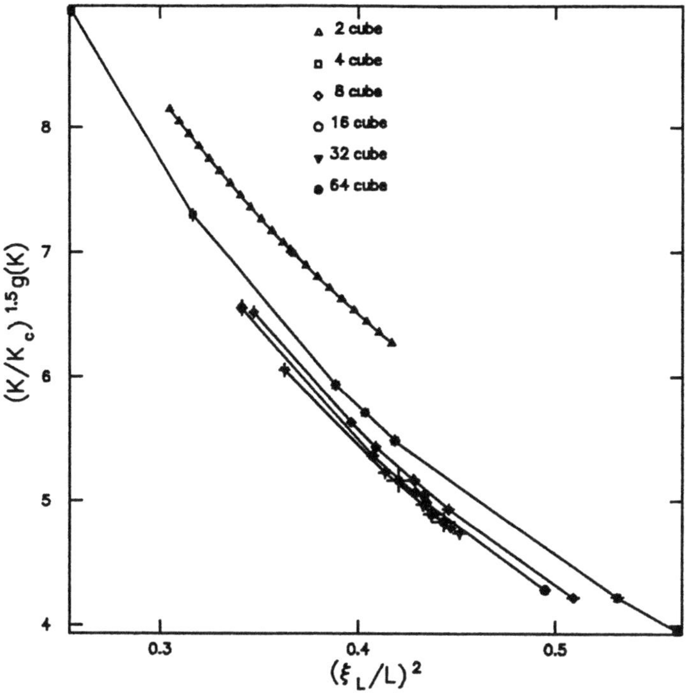

Fig. 1. Monte Carlo results for $g(K,L)(K/K_c)^{1.5}$ *vs.* $(\xi_L/L)^2$ for the three dimensional Ising model.

Kawashima [2] recently found that the second of these two limits can be taken as a lower bound on the first, and so the question of whether $g^* > 0$ or not could be answered, if one could show that the second limit $\hat{g} > 0$.

A key feature to notice in analysing the nature of $\hat{g}$ is the observation of Baker and Erpenbeck [3] that, for the three-dimensional, Ising model in the range where ξ_L/L is bounded away from zero, $(K/K_c)^{1.5}g(K,L)$ plotted against ξ_L/L converges fairly rapidly as $L \to \infty$ to a single function. We have done more calculations of these quantities for the three-dimensional, Ising model on the simple-cubic lattice, using the same computer code as in [2] which is a Swendsen-Wang [4] procedure with improved estimators. The results are displayed in Fig. 1. The reason for the factor of $(K/K_c)^{1.5}$ is to keep the quantity plotted finite as $K \to 0$, and unchanged at the critical point. We have used the estimate of Gupta and Tamayo [5], $K_c \approx 0.221655$ here. This value also agrees within error with the recent results of Blöte *et al.* [6], $K_c \approx 0.2216546 \pm 10$.

In the limit as the system size becomes large, the Binder [7] cumulant ratio, $U = \langle M^4 \rangle / (\langle M^2 \rangle)^2 - 3$, where M is the sum over the lattice of the Ising spins, is zero for $K < K_c$, minus two for $K > K_c$ and some intermediate, fixed-

Table 1. Table of the curve crossing values of K, g, $(\xi_L/L)^2$ and U.

$n \otimes 2n$	$K_{g,\otimes}$	$g_\otimes$	$K_{\xi,\otimes}$	$(\xi_L/L)^2_\otimes$	$K_{U,\otimes}$	$U_\otimes$
$2\otimes 4$	0.2135±2	8.24±3	0.2164±2	0.337±1	0.2274±3	-1.5836±7
$4\otimes 8$	0.2209±1	5.77±6	0.2213±2	0.411±6	0.2227±1	-1.501±4
$8\otimes 16$	0.22156±7	5.2±1	0.22161±4	0.426±5	0.22201±2	-1.489±2
$16\otimes 32$	0.22164±2	5.1±1	0.22165±1	0.432±3	0.22168±2	-1.425±9
$32\otimes 64$	0.22165±1	5.0±2	0.22165±2	0.428±10	0.22166±1	-1.414±7

point value U^* for $K = K_c$. Binder has argued that the crossover point of U for successively larger system sizes is a good estimator for the critical temperature. Baker [1] found this proceedure to be about an order of magnitude better than using the peak in the susceptibility. He also noted that the structure of the approach to the infinite, system-size limit is such that the renormalized coupling constant must also display a similar set of crossover points giving estimates of the critical temperature. For the purpose of computing the crossing it is useful to have systematically varying curves, such as used by Gupta and Tamayo [5]. They used a histogram reweighting scheme. To the same end, we have used the same sequence of random numbers for the various different temperatures on the same system size. In addition we have made a series of runs with independent sets of random numbers for the different sized systems at $K = 0.221655 \approx K_c$. The values of the crossings found for K, g, $(\xi_L/L)^2$, and U are listed in Table 1. The 2^3 results were done exactly, numerically. The 4^3, 8^3, and 16^3 were run 300,000 Monte Carlo sweeps or 400,000 for some of the 16^3 cases. The 32^3 and 64^3 were run 400,000 Monte Carlo sweeps, except for U for $L = 64$ the better data of Gupta and Tamayo [5] was used. Initialization was several hundred sweeps, with the number increasing with system size. These crossings have been determined by linear interpolation from our data. The quoted errors are single standard deviations and are in units of the last figure quoted. It is to be noticed that, within error, the crossing values for g and $(\xi_L/L)^2$ move to higher values of K, while those for U move to lower values giving confidence that we may have, within our error, bracketed the critical temperature. The results from U are not in contradiction with those of [5], where there is seen some possibility of non-monotonic behavior. The errors prevent us from making a clear distinction here. Thus our results bracket from above and below, within their errors, the value of K_c proposed in refs. [5 & 6]. The results for the crossing values for g are in contrast to the results in two dimensions [1], where the crossing value of K decreases with system size and the value of g at the crossing increases.

The error, $|K_{g,\otimes} - K_c|$, appears to be very roughly proportional to L^{-3} for the crossing values determined by $g(K)$. The rate that $g(K_{g,\otimes})$ approaches $\hat{g}$

Table 2. Table of finite size values for $K = 0.221655 \approx K_c$.

L	$U(K)$	$(\xi_L/L)^2$	$g(K)$	$g'(K)$	$g(K_{U,\otimes}, L)$
2	-1.55264	0.36584	7.01740	-140	
4	-1.483±2	0.418±2	5.49±4	-390	3.96±11
8	-1.445±2	0.428±2	5.16±4	-780	4.40±11
16	-1.421±3	0.430±2	5.05±3	-1900	4.26±7
32	-1.410±4	0.433±3	4.97±4	-6800	4.8±2
64	-1.407±4	0.437±4	4.89±6	-14000	4.9±2
128	-1.404±4				
256	-1.397±6				

is just a reflection in the curve of Fig. 1 of rate of approach of $K_{g,\otimes}$ to K_c. The rates of convergence of the other series are less clear.

We report in Table 2, in analogy with the presentation of ref. [1], the values taken at our estimated critical point as the system size increases. We are able to extend the system-size for U to $L = 128,\ 256$ by the use of the data of ref. [5]. The cross comparison of this data with the current calculations for $L = 64$ is within the error bars. The same extension can not be made for the other quantities as the extrapolation of the estimator of ξ_L to zero momentum was not incorporated in the analysis of [5]. This difference is irrelevant for the thermodynamic limit, g^*, but leads to about a 10% difference in $\hat{g}$. It is to be noted that values of g, $(\xi_L/L)^2$ and U appear to converge rapidly as the system size increases, and the system-size variation disappears into the error near the bottom of the table. The picture of $g(K)$ presented in refs. [1 & 2] envisions a slope which rapidly increases with system size at the critical point to accommodate $g^* \neq \hat{g}$. This view is well supported by the (very rough) values of $g'(K)$ which were obtained by numerical differentiation and are displayed in Table 2. These values increase roughly proportional to $L^{4/3}$. A more detailed study of the slopes of other quantities was given in ref. [5].

In the last column of Table 2, we give the values of $g(K, L)$ at the crossing point of $U(K, L)$ and $U(K, \frac{1}{2}L)$. As remarked above, this crossing point seems to be an upper bound for K_c. As g is a decreasing function of K, this quantity should be a lower bound for $\hat{g}$. We note that $g(K_{U,\otimes}, L)$ as just defined and tabulated in Table 2, is a generally increasing function of the system size, L. There is a competition reflected here between the increase in the slope of g and shown in the previous column which tends to lower the value, and the rate of convergence of $K_{U,\otimes}$ which tends to increase the value. As the latter seems to be dominate, asmptotically, we have thus constructed what appears to be numerically, an asymptotically increasing lower bound to $\hat{g}$. Hence, from the values in Table 2, we conclude at the 2 standard error level that $g^* > \hat{g} \gtrsim 4.5 > 0$. This more through study bolsters the conclusion of ref. [2], that $g^* > 0$ and so that hyperscaling holds. Note that the same type arguments applied to the results in Table 1, indicate further likely bounds. If we use two standard errors then we find $5.4 \gtrsim \hat{g}$. Similarly we find a lower bound for the large system limit of $(\xi_L/L)^2$ of 0.426, and $U^* \gtrsim -1.43$.

This work has been supported in part by the U.S. D.O.E. The authors are happy to acknowledge helpful discussions with J. J. Erpenbeck.

References

1. G. A. Baker, Jr., J. Stat. Phys. **77**, 955 (1994).
2. G. A. Baker, Jr. and N. Kasashima, Phys. Rev. Letts. **75**, 994 (1995).
3. G. A. Baker, Jr. and J. J. Erpenbeck, *Computer Simulation Studies in Condensed Matter Physics* **7**, eds. D. P. Landau, K.K. Mon and H.-B. Schüttler, ((Springer, New York, 1994) pg. 213.
4. R. H. Swendsen and J. S. Wang, Phys. Rev. Letts. **58**, 86 (1987).
5. R. Gupta and P. Tamayo, to be published, J. Mod. Phys. C (1996).
6. H. W. J. Blöte, E. Luijten, and J. R. Heringa, Poster presented at the Nineteenth IUPAP International Conference of Statistical Physics, Xiamen, China, 31 July - 4 August, 1995 & cond-mat/9509016.
7. K. Binder, Z. Phys. B **43**, 119 (1981).

Phase Transition in the Two-Dimesional Classical XY Model

J.K. Kim

Center for Simulational Physics and Department of Physics,
The University of Georgia, Athens, GA 30602, USA

For the two dimensional classical XY model we present extensive high - temperature -phase bulk data extracted based on a novel finite size scaling (FSS) Monte Carlo (MC) technique, along with FSS data near criticality. Our data verify that $\eta = 1/4$ sets in near criticality and clarify the nature of correction to the leading scaling behavior. It is pointed out that the result of conventional FSS analysis near criticality is inconsistent with other predictions of Kosterlitz's renormalization group approach.

The mechanism of the phase transition in the two dimensional classical XY model was first illustrated by Berezinskii, and by Kosterlitz and Thouless[1] based on vortex binding scenario. The prediction is that in the high temperature phase the critical behaviors of correlation length (ξ) and magnetic susceptibility (χ) are given by

$$\xi(t) \sim \exp(bt^{-\nu}), \quad \chi(t) \sim \xi(t)^{2-\eta}. \tag{1}$$

Here $t = (T - T_c)/T_c$, and the predicted value of ν and η are 1/2 and 1/4 respectively. Corrections to the expressions in Eq.(2) were calculated[2]

$$\xi(t) \sim \exp(bt^{-1/2}) \, [1 + \mathcal{O}(t)], \quad \chi(t) \sim t^{-1/16} \exp(\frac{7}{4}bt^{-1/2}). \tag{2}$$

Accordingly $\chi(t)/\xi(t)^{7/4}$ increases as $t \to 0$.

Most previous MC studies[3] and series expansion methods[4, 5] have determined T_c by assuming $\nu = 1/2$ in Eq.(1). In this fashion $0.890 \leq T_c \leq 0.894$ was generally obtained. The generic feature emerged from those MC studies was that the values of η calculated from the high temperature phase thermodynamic value (value in the thermodynamic limit) of ξ and χ are significantly larger than the predicted value.

In this work we present extensive MC data of the thermodynamic values of ξ and χ up to T=0.93 ($\xi \simeq 1391(22)$) that are obtained using a novel FSS technique[6, 7, 8, 9]. The technique is based on a fundamental formula of FSS[6],

$$A_L(t) = A_\infty(t) Q_A(x), \quad x = \xi_L(t)/L. \tag{3}$$

Here A_L denotes a multiplicative renormalizable quantity A defined on a finite lattice of linear size L, and Q_A is a universal scaling function of a different scaling variable $\xi_L(t)/L$. Eq.(3) is supposed to hold even for the 2D XY model[10]. For a detailed description of the technique see Refs.[6, 7, 8], where the thermodynamic

Springer Proceedings in Physics, Volume 82
Computer Simulation Studies in Condensed-Matter Physics IX
Eds.: D. P. Landau, K.K. Mon, H. -B. Schüttler

Table 1: The thermodynamic values of ξ and χ for $0.93 \leq T \leq 0.98$, extracted through our MC FSS technique.

T	0.98	0.97	0.96	0.95	0.94	0.93
ξ	70.4(4)	100.3(7)	155.7(1.5)	262.7(2.9)	539.3(4.5)	1391(22)
χ	4284(20)	7932(30)	16978(55)	42295(108)	147536(614)	$7.70(9)\times10^5$

values extracted by employing the technique are shown to agree completely with those measured traditionally for the 2D and 3D Ising models[8], 2D XY model[6], and 2D Heisenberg ferromagnet[7, 9].

In this work the extractions of the thermodynamic values of ξ and χ are made in the range $0.93 \leq T \leq 0.98$, based on available data of $(x, \mathcal{Q}_A(x))$ for $T = 1.0$ [6]. The values of L over the range $80 \leq L \leq 480$ are used for our analysis, and most interpolations are made in the range $\mathcal{Q}_A(x) \geq 0.1$. We employed Wolff's single- cluster algorithm[11] for the MC simulations, imposing periodic boundary condition on square lattice. For a given set of T and L typically 20 - 40 numbers of bins have been obtained for the measurements, with each bin being composed of 10 000 measurements each of which is separated by 3-12 consecutive one cluster updating.

Our thermodynamic data, tabulated in Table (1), agree remarkably well with those that become available recently from a strong-coupling analysis[12]. We carried out a χ^2 fit for our thermodynamic data of ξ over the range $5.01(3) \leq \xi \leq 1391(22)$ (corresponding to $0.93 \leq T \leq 1.19$), assuming the exponential singularity Eq.(1). The best fit is obtained for

$$T_c \simeq 0.893(3) \quad \text{and} \quad \nu \simeq 0.48(4), \tag{4}$$

with $\chi^2/N_{DF} \simeq 0.75$. Our estimate of ν agrees remarkably well with Eq.(1). $\nu = 0.5$ holds at $T_c \simeq 0.892$ with $\chi^2/N_{DF} \simeq 0.80$, being in agreement with the range of T_c conventionally accepted.

Fig.(1) depicts $\chi(T)/\xi(T)^{2-\eta}$ with $\eta = 1/4$ as a function of $\ln(\xi)$. It shows that the effective value of η gradually decreases to $\eta = 1/4$ as $T \to T_c$. It turns out that the data fit perfectly well to

$$\chi(t)/\xi(t)^{7/4} = a + b|\ln(t)|^r, \tag{5}$$

with $a \simeq 1.833$, $b \simeq 0.955$, $r \simeq -0.413$, and $T_c = 0.893$ ($\chi^2/N_{DF} < 0.1$). We remark that the slow decrement of $\chi(T)/\xi(T)^{2-\eta}$ as $T \to T_c$ in Fig.(1), is inconsistent with any positive value of r such as predicted in [2]. Without taking account of the correction to Eq.(1), on the other hand, the value of η turns out to be 0.272(8) with $\chi^2/N_{DF} \simeq 9.3$. Accordingly, correction to scaling in Eq.(1) is essential.

We employ a conventional FSS technique to locate T_c independently, i.e., without any assumption of the functional form of critical singularity. Namely, we measured both χ_L and the 4th order cumulant ratio[13] defined as $U_L \equiv$

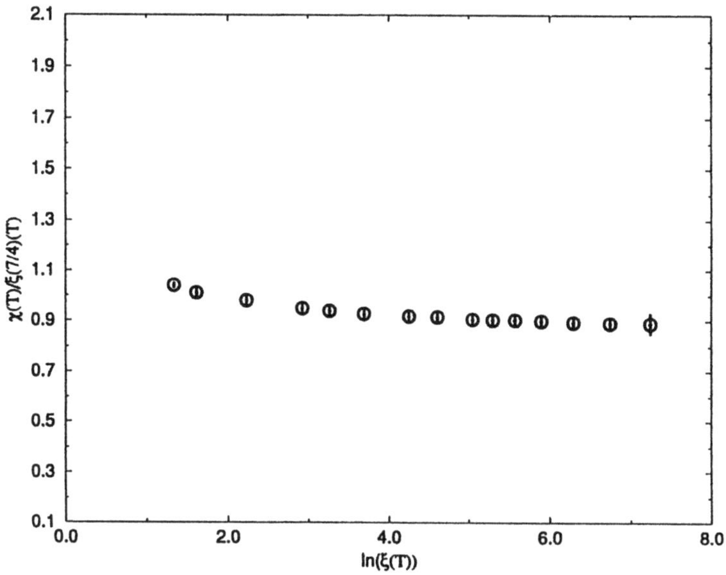

Fig.(1): $\chi(T)/\xi^{7/4}(T)$ versus $\ln(\xi(T))$ for T over the range, $0.93 \leq T \leq 1.25$. The thermodynamic data for $T > 0.98$ are taken from Ref.[6]. The tendency to decrease implies $\eta > 0.25$, but the values of $\chi/\xi^{7/4}$ tend to stablize at approximately $\ln(\xi) \geq 5$.

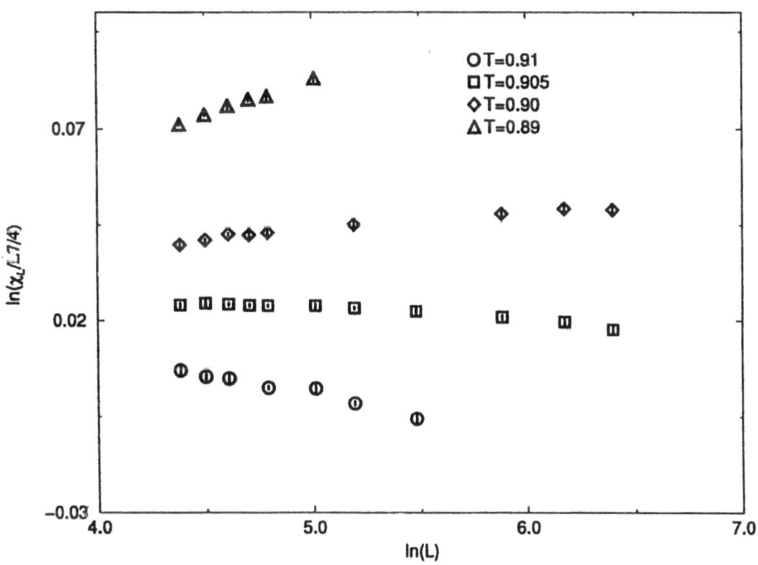

Fig.(2): $\ln(\chi_L/L^{7/4})$ versus $\ln(L)$ over $80 \leq L \leq 600$. Here the slope of a straight line is related to the value of η, that is, $\eta = 1/4 - slope$. Note a slow decrease (increase) of $\ln(\chi_L/L^{7/4})$ with respect to $\ln(L)$ for T=0.905 (T=0.90); hence, $0.90 < T_c < 0.905$ with $\eta = 1/4$.

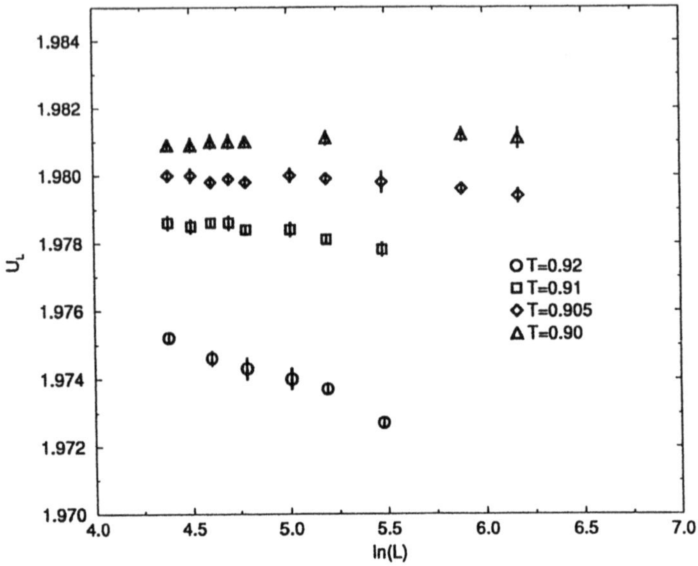

Fig.(3): U_L as a function of $\ln(L)$ over $80 \leq L \leq 600$, for various temperatures near criticality. U_L is L independent (within small statistical error) over $80 \leq L \leq 150$, and $80 \leq L \leq 360$ for $T = 0.91$ and 0.905 respectively (pseudo scale invariance), although it eventually decreases with L.

$3- < S^4 > / < S^2 >^2$ with $S^2 \equiv |\sum_i \vec{s}_i|^2$, by varying L from 20 to 600 for some temperatures over the range $0.89 \leq T \leq 0.92$. Certain *non-asymptotic* FSS behavior was observed for $L < 80$, which seems to be a consequence of the correction to Eq.(3).

Our data with $L \geq 80$ fit very well to a conventional FSS formula at criticality $\chi_L \sim L^{2-\eta}$ for $T \leq 0.905$ ($\chi^2/N_{DF} < 0.5$) (Fig.(2)), suggesting that the system remain critical over the range of temperature. The extracted values of η from the linear fit are: $\eta(T) = 0.260(1)$, $0.252(0)$, $0.245(0)$, and $0.231(1)$ for T=0.91, 0.905, 0.90, and 0.89 respectively. Therefore, $\eta(T_c) = 0.25$ is consistent with $0.90 \leq T_c \leq 0.905$, most probably $T_c \simeq 0.904$. This estimate is also consistent with our data of U_L (Fig.(3)). $U_L(T = 0.92)$ definitely decreases with L from L=80 to L=240 so that T_c is obviously smaller than 0.92. At T=0.91 and 0.905 it varies so mildly that some fictitious scale invariance of U_L appears over a range of L not too large; nevertheless, it eventually decreases.

Notice, however, that the estimate of T_c based on the conventional FSS analysis is significantly larger than that estimated by assuming the exponential singularity Eq.(1). Assuming $T_c = 0.904$ in Eq.(1) the best fit is for $\nu \simeq 0.30(1)$ with $\chi^2/N_{DF} \simeq 6.7$; assuming both $T_c = 0.904$ and $\nu = 0.5$ the best fit has $\chi^2/N_{DF} \simeq 263$. Therefore, $\nu = 1/2$ in Eq.(1) is definitely inconsistent with the prediction of the conventional FSS at criticality. We also checked whether or not the data fit to $\xi = c_1(1 + c_2 t^\theta)\exp(bt^{-\nu})$ with the input value of T_c over

$0.90 \leq T_c \leq 0.905$. It turns out that they fit to the formula for some *negative* value of θ only with the value of χ^2/N_{DF} being almost insensitive to the choice of T_c. However, this function has too much freedom to the fit and gives unstable predictions for the values of both T_c and ν.

On the other hand, the thermodynamic data fit very well to a modified second order phase transition, $\xi = c_1(1 + c_2 t^\theta)t^{-\nu}$ Assuming $T_c = 0.904$ we would estimate that $\nu \simeq 3.10(8)$, $\theta \simeq 1.82(8)$, and $c_2 \simeq 44.2$ with $\chi^2/N_{DF} \simeq 1.2$.

To conclude: We have reported strong numerical evidence that $\eta = 1/4$ sets in at temperatures close to criticality. We have been able to verify Kosterlitz's exponential singularity based on *unconstrained* fit of our high temperature thermodynamic data. In spite of the extraordinary broadness of our high temperature data, the possible existence of an ordinary second order phase transition still remains, so a knowledge of *true* critical temperature is crucial to the resolution of the order of the phase transition in the 2D classical XY model. The nature of the correction to the leading scaling behavior Eq.(1) has been clarified. Prediction of T_c based on conventional FSS analysis used at criticality is shown to be incompatible with the conventional exponential singularity.

After the completion of the current work we have become aware of a recent related work[14] that questions, based on a FSS study of Lee-Yang zeros, the vortex binding scenario in the 2D XY model: The authors argue that Eq.(1) cannot be compatible with $\eta = 1/4$ without a certain correction. However, the proposed correction, $r \simeq 0.02$ in Eq.(5), is *inconsistent* with our picture in Fig.(1) in that their $\chi(t)/\xi^{7/4}(t)$ *increases* as $t \to 0$.

Current work was partially supported by NSF grant DMR-9405018.

References

[1] V. L. Berezinskii, Zh. Eksp. Teor. Fiz. **59** (1970) 907 [Sov. Phys. JETP **32** (1971) 493]; J. M. Kosterlitz and D. J. Thouless, J. Phys. C **6** (1973) 1181; J. M. Kosterlitz, J. Phys. C **7**(1974) 1046

[2] D. J. Amit, Y. Y. Goldschmidt and G. Grinstein, J. Phys. A **13** (1980) 585; L. P. Kadanoff and A. B. Zisook, Nucl. Phys. B **180** (1981) 61

[3] U. Wolff, Nucl. Phys. B **322** (1989) 759; R. G. Edwards, J. Goodman and A. D. Sokal, Nucl. Phys. B **354** (1991) 289; R. Gupta and C. F. Baillie, Phys. Rev. B **45** (1992) 2883; W. Janke and K. Nather, Phys. Rev. B **48** (1993) 7419; P. Olsson, Phys. Rev. Lett. **73** (1994) 3339; M. Hasenbusch, Marcu and K. Pinn, Physica A **208** (1994) 124

[4] P. Butera and M. Comi, Phys. Rev. B **47** (1993) 11969

[5] M. Campostrini, A. Pelissetto, P. Rossi and E. Vicari, Heplat/9603002 (1996)

[6] J.-K. Kim, Europhys. Lett **28** (1994) 211

[7] J.-K. Kim, Nucl. Phys. B (Proc. Suppl.) **34** (1994) 702; *ibid*, Phys. Rev. D **50** (1994) 4663

[8] J.-K. Kim, A. J. F. de Souza and D. P. Landau, Phys. Rev. E (1996) (to appear)

[9] S. Caracciolo, R. G. Edwards, A. Pelissetto and A. D. Sokal, Phys. Rev. Lett. **74** (1995) 2969; *ibid*, **75** (1995) 1891

[10] E. Brézin and J. Zinn-Justin, Nucl. Phys. B **257** [FS14] (1985) 867

[11] U. Wolff, Phys. Rev. Lett. **62** (1989) 361

[12] The magnetic susceptibility and correlation length obtained from strong coupling expansion are compared with our Monte Carlo data in Ref.[5].

[13] K. Binder, Z. Phys. **43** (1981) 119

[14] R. Kenna and A.C. Irving, Heplat/9601029 (1996)

Cluster Structures in the Configuration Space and Relaxation in the $3d \pm J$ Ising Spin Glass

T. Klotz

Technische Universität Dresden, Institut für Theoretische Physik, D-01062 Dresden, Germany

Abstract. The relaxation behaviour of a small cubic $\pm J$ spin glass is considered. The long-time relaxation is only determined by the largest non-trivial eigenvalues of the transition matrix of this system. Based on the knowledge of the cluster structure in the configuration space for energetically low-lying states two approximations for the transition matrix are proposed.

The complicated structure of the phase space is the key to understand the unusual relaxation and aging behaviour of spin glasses. Various attempts have been made to describe this structure. In this contribution the exact knowledge of the global microscopic low-energy configuration-space structure is used to study the relaxation of the system by random walks in the configuration space [1]. For long times and low temperatures such a random walk is only determined by the largest nontrivial eigenvalues and eigenvectors of the temperature-dependent transition matrix of the energetically low-lying states of the system [1, 2, 3]. Therefore, it is more convenient to calculate these eigenvalues and eigenvectors directly. This can be done exactly by solving the eigenvalue problem of the very large and sparse transition matrix numerically.

On the other hand, the long-time behaviour of the system should not depend on all microscopic details of the configuration space structure. Thus, it seems to be useful to simplify the complete transition matrix by using the knowledge of the main features of the energy landscape. A good model of the transition matrix should reproduce at least the typical behaviour of the largest nontrivial eigenvalues. Two models will be presented and discussed in this paper.

The method of branch-and-bound in a recursive version [4] was used to find the ground states of a cubic Ising spin glass system on a $4 \times 4 \times 4$ lattice with $\pm J$ interactions between nearest neighbours and periodical boundary conditions in all three directions. The distribution of interactions was chosen randomly with an exact fraction of 50 % of ferromagnetic and of antiferromagnetic bonds.

In addition to the ground states, all M energetically low-lying states up to the second excitation were calculated ($M \sim 10^5$). For a representative system of 1000 investigated samples all these states have been analyzed with respect to their neighbouring relations. The found complex structure in the configuration space can be visualized in a simple schematic [5] or in a more detailed picture [1]. Even for this small system size, all the suggested basic ingredients of the configuration space as valleys, clusters, energy barriers and the connections between different components can be found.

Springer Proceedings in Physics, Volume 82
Computer Simulation Studies in Condensed-Matter Physics IX
Eds.: D. P. Landau, K.K. Mon, H. -B. Schüttler

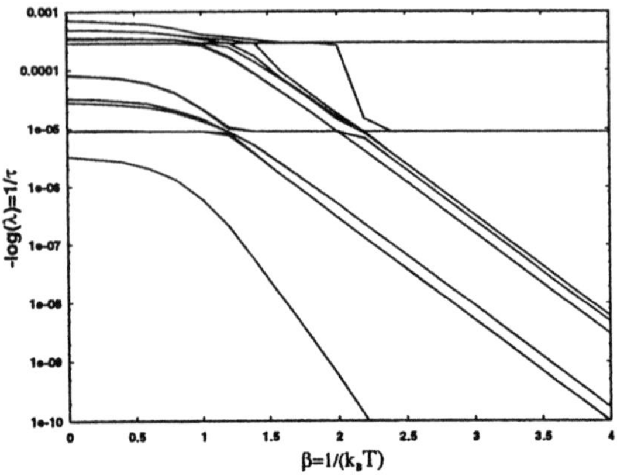

Figure 1: Largest non-trivial eigenvalues for a representative system.

The transition matrix of such a structure is given by the probabilities w_{ij} for a transition from state j to state i during one time step. Considering only single-spin flips, all transition probabilities vanish between states, which are not neighbours in the configuration space. The remaining elements of the transition matrix are just determined by the energy difference between the two states ($w_{ij} \propto (1/N) \min\{\exp[-\beta(E_i - E_j)], 1\}$). This leads to a very large and sparse non-symmetric matrix. Only in the case of infinite temperature it is a symmetric one. But for finite temperatures the matrix can be transformed into a symmetric one using the condition of detailed balancing in the equilibrium [2]. Fig. 1 shows the largest eigenvalues for the transition matrix of a representative system taking into consideration all connected states up to the second excitation [6].

For low temperatures the phase space breaks up into different metastable valleys of the configuration space. Therefore, the transition matrix will be separated into independent blocks, too. This leads to a diverging relaxation time τ for each metastable valley. The slope for the inverse relaxation time in Fig. 1 is just determined by the height of the energy barriers between different metastable valleys. The other eigenvalues in Fig. 1 should correspond to the relaxation processes, which do not overcome energetic barriers of the system.

The simplest approximation for the found structure in the configuration space is given by a fully unstructured system as discussed in [1]. There are no energetic barriers in such a trivial model. Therefore, the eigenvalues are independent on the temperature. But even in the case of infinite temperature, which corresponds to a flat phase space, the quantitative deviation to the really found relaxation times is very strong. Obviously, not only the energetic structures given by the energy barriers, but the entropical barriers given by the inner-cluster structures are important for a description of the relaxation processes.

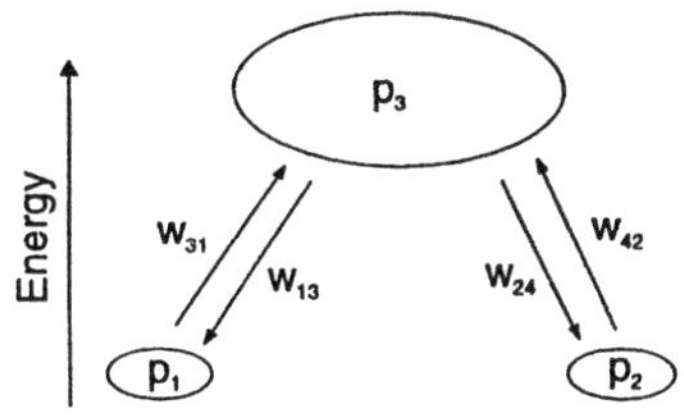

Figure 2: Schematic picture of model A. (The p_i denote the total probabilities in the different clusters.)

The basic elements of the found configuration space structure are clusters, which consists of states of the same energy. Thus the relaxation inside of such a cluster is not influenced by the temperature. The only transition probabilities, which are influenced by the temperature, describe transitions between clusters of different energy. Therefore, a nontrivial model of the configuration space should at least include the transitions between different clusters and a more or less raw picture of a single cluster.

The transition probability between two different clusters can simply be approximated by counting the number of connections between these clusters and the number of states in the boundary regions of these clusters. The first and simplest approximation for a cluster should be to forget the inner structure and just to 'collapse' the cluster into one state. This is done in model A (Fig. 2).

In order to compare the model to the 'real' system, we take a typical subset of three large clusters of the total structure of the configuration space, which form an energetic barrier. On the one hand, the largest nontrivial eigenvalues are calculated directly using the full transition matrix of this subset. On the other hand we approximate these values using the model. The qualitative behaviour of the eigenvalues of model A is in good agreement to the real system (see Fig. 4), but quantitatively the relaxation process in the model is too fast. This seems to be caused by the underestimated influence of the relaxation inside of the clusters.

For low temperatures the most probable states should be concentrated in and around the energetically low-lying states. For higher temperatures the inner part of the clusters should become more important.

Therefore, in the improved model B the upper cluster was splitted up into the two boundary regions, which connect this cluster with its neighbouring clusters, and a fictive rest cluster. To describe the main relaxation processes inside of the complete cluster as good as possible, this rest cluster should be far away from the boundary regions. This can be realized by calculating the components of the boundary regions in relation to the largest eigenvectors of the isolated transition matrix of the considered cluster. The fictive rest cluster should have complementary components to these boundary regions. The transition probabilities between the three sub-clusters can easily be calculated using again the temperatur-independent eigenvalues and eigenvectors of the transition matrix for this cluster.

The two energetically low-lying clusters are much smaller than the upper cluster and all the states inside of these clusters belong to their boundary regions. Therefore, only the upper cluster was splitted.

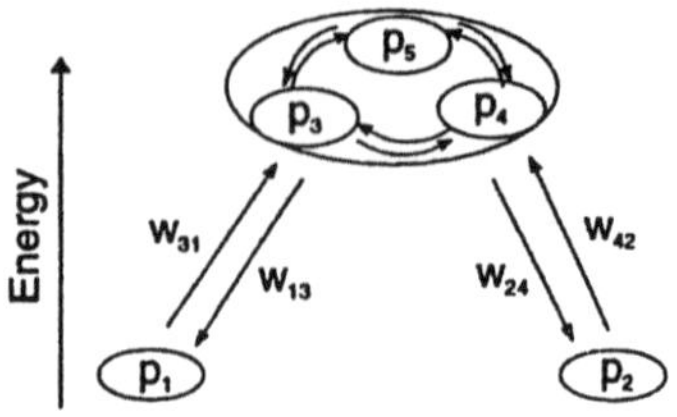

Figure 3: Schematic picture of model B. (The p_i denote the total probabilities in the different clusters.)

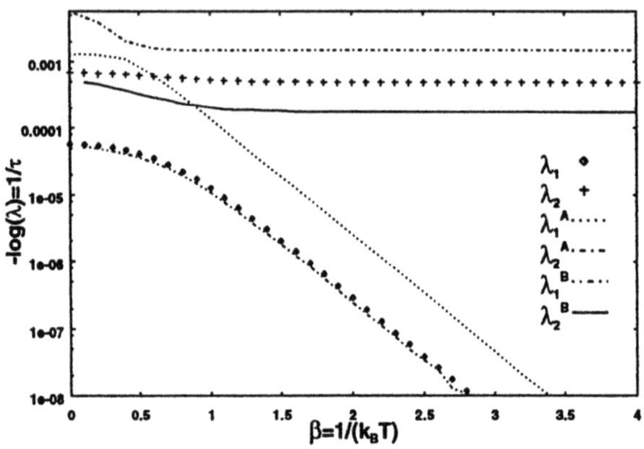

Figure 4: Largest non-trivial eigenvalues λ_1 and λ_2. (The crosses and diamonds are results of the direct calculation. The lines belong to model A and B.)

Fig. 3 shows the scheme of model B. The comparison of the calculated eigenvalues to the exact values shown in Fig. 4 gives an excellent agreement at least for the most important eigenvalue λ_1 for all temperature region.

Thus, by a very raw model of the exact microscopic energy landscape it is possible to reduce the size of the transition matrix drastically at minimal loss of information for the largest nontrivial eigenvalues, which describe the physical interesting long-time relaxation behaviour of the system.

Acknowledgement. The author wish to thank S. Kobe for collaboration and fruitful discussion. This work was supported by the Deutsche Forschungsgemeinschaft (project no. Ko 1416).

References

[1] T. Klotz and S. Kobe, *Acta Phys. Slovaca* **44** (1994) 347.

[2] K. Kawasaki, *Phys. Rev.* **148** (1966) 375.

[3] H.-U. Bauer, K. Schulten and W. Nadler, *Phys. Rev.* **B38** (1988) 445.

[4] A. Hartwig, F. Daske and S. Kobe, *Computer Phys. Commun.* **32** (1984) 133.

[5] T. Klotz and S. Kobe, *J. Phys. A: Math. Gen.* **27** (1994) L95.

[6] T. Klotz and S. Kobe, in: *Hayashibara Forum '95, Int. Symp. Coherent Approaches to Fluctuations, Kyoto 1995* (M. Suzuki, N. Kawashima, eds.), World Scientific, Singapore 1996, p. 192.

Application of Combinatorial Optimization to the Problem of Configuration Space of Ising Spin Glasses

S. Kobe

Technische Universität Dresden, Institut für Theoretische Physik, D-01062 Dresden, Germany

Abstract. Methods of combinatorial optimization can be applied to find the exact ground state and the exact low-energy configuration space of spin glass models. The branch-and-bound algorithm is used to show that exact results for finite systems can be obtained although not all spin configurations are enumerated.

The determination of the exact ground state with the energy E_0 (in dimensionless units) of the Ising model ($s_i = \pm 1$) with N spins and arbitrary exchange coupling constants J_{ij} of both signs,

$$E_0 = \min_{s_i=1\vee-1} \left(-\sum_{i<j}^{N} J_{ij}\, s_i\, s_j \right) , \tag{1}$$

belongs to the hard problems of discrete nonlinear optimization. Because the number of states increases with 2^N, the solution of (1) by complete enumeration is restricted to very small systems. Methods of combinatorial optimization allow to find the exact solution avoiding the calculation of states with higher energy values. This is shown in [1] for an Ising model with frustration using the branch-and-bound algorithm, see also [2]. Other optimization methods are applied to the $\pm J$ Ising model, e.g. by Bieche et al. [3] and Barahona et al. [4] using the perfect matching of minimum weight in the graph of frustrated plaquettes. Ideas of polyhedral combinatorics are used by Grötschel et al. [5] and De Simone et al. [6].

The general principle of the branch-and-bound algorithm can be easily demonstrated considering a simple example. It concerns a cluster with $N = 7$ spins and antiferromagnetic couplings of different strength. The upper triangle of the interaction matrix $\mathbf{J} = (J_{ij})$ is given by

$$\mathbf{J} = \begin{pmatrix} 0 & -5 & -2 & -5 & -6 & -1 & 0 \\ & 0 & -10 & -4 & 0 & -2 & -1 \\ & & 0 & 0 & 0 & -3 & 0 \\ & & & 0 & -3 & -5 & -7 \\ & & & & 0 & -4 & -5 \\ & & & & & 0 & 0 \\ & & & & & & 0 \end{pmatrix} . \tag{2}$$

Springer Proceedings in Physics, Volume 82
Computer Simulation Studies in Condensed-Matter Physics IX
Eds.: D. P. Landau, K.K. Mon, H. -B. Schüttler

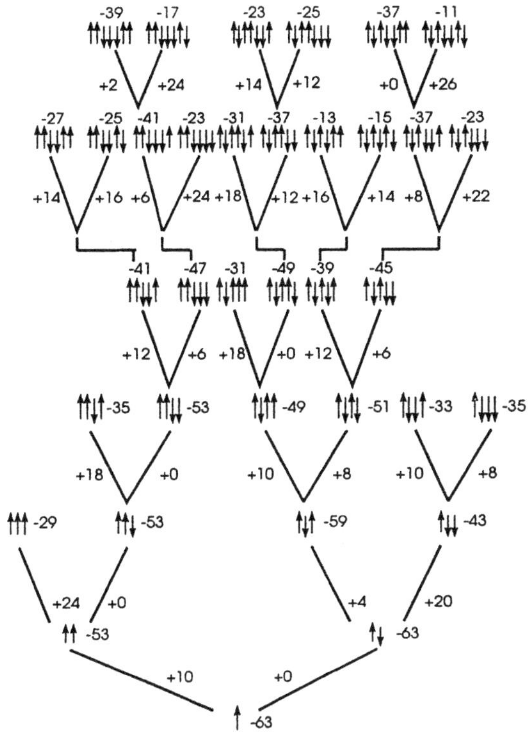

Fig. 1. Branch-and-bound tree for a cluster with $N = 7$ spins given by the interaction matrix (2); $E_{id} = -63$, $E_{cut} = -37$, $E_0 = -39$.

In Fig. 1 a tree is constructed by putting the spins successively in fixed directions. When the spin number l is fixed, the branching level l is indicated and at each branching node a configuration and an energy value E_l is given. So for $l = 1$ the spin number 1 is fixed ("up") and for $E_1 = E_{id}$ is chosen, where $E_{id} = -\sum |J_{ij}|$ is a lower bound for the ground-state energy E_0 in Eq. (1). The other values E_l can be obtained by the following rule ($l > 1$):

$$E_l = E_{l-1} + 2 \sum_{k(||l)}^{l-1} |J_{kl}|. \tag{3}$$

Here $k(||l)$ denotes a summation over such k only, for which the spins had been fixed already in the same direction as the spin l. In other words, the sum contains only such contributions, which arise due to "frustration" of the spin l with all spins fixed before. (It should be noted that this concept has to be followed, if models with mixed signs of J_{ij} are considered instead of purely negative ones as it is chosen in example (2)). In Fig. 1 the values of the second term on the right-hand-side of (2) are given at the branching lines. From Eq.

(3) it follows

$$E_l \geq E_{l-1}. \tag{4}$$

It it easy to recognize that only states and energies on the top of the tree (i.e. $l = N$) belong to the system (2).

A good heuristic approximation (even though not the solution of (1)) can be obtained following the connection between the starting point E_1 and E_N according the deepest descent. In our example the energy of this state is -37. This energy value is used as E_{cut} to bound the branchings at all nodes with $E_l > E_{cut}$. It is guaranteed by Eq. (4) that all cutted branches could only lead to states $E_N \geq E_{cut}$ and therefore none of these 'lost' states can be a solution of the problem (1). So the resulting tree in Fig. 1 can be used for a search strategy for E_0 instead of the complete enumeration of all states. The numerical effort is estimated at $2^{\alpha N}$ with $\alpha = 0.27$ for the determination of all ground states of cubic $\pm J$ spin glasses.

Moreover, it is obvious that this algorithm also yields *all* low-lying states with energies lower than E_{cut}, if $E_{cut} > E_0$ is chosen.

It is worthwhile to consider a system without frustration following the proceedure of 'branch-and-bound'. In this case the ground-state energy can be found already by the deepest descent and E_0 is indentical with E_{id}. Consequently, the difference $E_0 - E_{id}$ is used in [7] to introduce a global measure of the frustration of the system.

The proposed algorithm is used to analyze the exact low-energy landscape in the configuration space of a $\pm J$ spin glass on a cubic 4^3 lattice with periodical boundary conditions in [8]. About 10^5 states, i.e. all states up to the second excitation, are calculated and ordered with respect to their connectivity. Starting from the complete knowledge of the configuration space the relaxation behaviour can be investigated, see also [9].

In summary, it is shown that methods of combinatorial optimization are suitable to obtain exact results for ground-state calculations of Ising spin glasses.

Acknowledgement. The author wish to thank A. Hartwig, F. Daske and T. Klotz for collaboration. This work is supported by the Deutsche Forschungsgemeinschaft (DFG project Ko 1416).

References

[1] S. Kobe and A. Hartwig, *Computer Phys. Communications* **16** (1978) 1.

[2] A. Hartwig, F. Daske, and S. Kobe, *Computer Phys. Communications* **32** (1984) 133.

[3] I. Bieche, R. Maynard, R. Rammal, and J. P. Uhry, *J. Phys. A: Math. Gen.* **13** (1980) 2553.

[4] F. Barahona, R. Maynard, R. Rammal, and J. P. Uhry, *J. Phys. A: Math. Gen.* **15** (1982) 673.

[5] M. Grötschel, M. Jünger, and G. Reinelt, in: *Heidelberg Colloquium on Glassy Dynamics, Lecture Notes in Physics, Vol. 275* (J. L. van Hemmen and I. Morgenstern, eds.), Springer, Berlin 1987, p. 325.

[6] C. De Simone, M. Diehl, M. Jünger, P. Mutzel, G. Reinelt, and G. Rinaldi, *J. Stat. Phys.* **80** (1995) 487.

[7] S. Kobe and T. Klotz, *Phys. Rev. E* **52** (1995) 5660.

[8] T. Klotz and S. Kobe, *Acta Phys. Slovaca* **44** (1994) 347; T. Klotz and S. Kobe, in: *Hayashibara Forum '95, Int. Symp. Coherent Approaches to Fluctuations, Kyoto 1995* (M. Suzuki, N. Kawashima, eds.), World Scientific, Singapore 1996, p. 192.

[9] T. Klotz, this volume.

Low-Temperature Metastable Lifetimes of the Square-Lattice Ising Ferromagnet

M.A. Novotny

Supercomputer Computations Research Institute, Florida State University, Tallahassee, FL 32306-4052 USA, and
Department of Electrical Engeneering, 2525 Pottsdamer Street, Florida A&M University-Florida State University, Tallahassee, FL 32310-6046, USA
e-mail: novotny@scri.fsu.edu

Abstract. We compare computer calculations of the lifetimes of the metastable state of the square-lattice Ising model at low temperatures with exact predictions. The computer simulations are obtained using the MCAMC algorithm, which combines Monte Carlo with absorbing Markov chains (MCAMC). Additional exact calculations using absorbing Markov chain matrices are also reported. It is found that for most values of the applied magnetic field there is a temperature-independent prefactor which must be included in order to compare theoretical predictions with the calculated lifetimes.

1. Introduction

The concept of the metastable state is of great interest since it applies to a number of ideas ranging from lifetimes of supercooled water to cosmological models. In addition, the ideas of droplet nucleation and growth have been applied to the switching behavior of nanoscale ferromagnets of relevance in magnetic recording media [1,2]. There have recently appeared a number of interesting new theorems that relate to the lifetime of Ising-like models at low temperatures. In particular, for the isotropic square lattice Jordão Neves and Schonmann [3] gave the exact equation for the lifetime as the temperature $T \rightarrow 0$ for fixed applied field H and fixed lattice size. One interesting prediction that has been made by later papers on the $T \rightarrow 0$ limit of metastability is that at low temperatures the nucleating droplet is not given by the droplet formed using the Wulff construction [4,5]. However, in all these cases one has to ask how low in temperature one needs to go in order to see the predicted results.

In this paper we concentrate on the isotropic square-lattice Ising nearest-neighbor (nn) ferromagnetic model in an applied magnetic field. The Hamiltonian is given by $\mathcal{H} = -J \sum_{\langle i,j \rangle} \sigma_i \sigma_j - H \sum_i \sigma_i$ where $\sigma = \pm 1$ and the first sum is over nn spins. The lattice is $L \times L$, and periodic boundary conditions are used.

The dynamic algorithm that will be used is: 1) with equal probability randomly pick one of the L^2 spins; 2) flip the chosen spin (i.e. let $\sigma_i \rightarrow -\sigma_i$) with a probability given by p_{flip}. We set Boltzmann's constant to unity. We choose $p_{\mathrm{flip}} = \min\left[\exp\left(\Delta E/T\right), 1\right]$, where ΔE is the energy difference between

Springer Proceedings in Physics, Volume 82
Computer Simulation Studies in Condensed-Matter Physics IX
Eds.: D. P. Landau, K.K. Mon, H. -B. Schüttler

the current spin configuration and the spin configuration with the chosen spin flipped. This is the traditional Metropolis dynamic [6].

In order to study metastability, in the standard fashion we start with the configuration with all spins $\sigma_i=1$ and apply a magnetic field $H<0$. We then monitor the number of spin-flip attempts required to go to a given cutoff magnetization. The average number of spin-flip attempts to the cutoff magnetization gives the lifetime τ of the metastable state. (To simplify comparison with the predicted values, we measure the lifetime in units of mcs — Monte Carlo steps, rather than the more traditional Monte Carlo steps per spin.) The predicted lifetime as $T\to 0$ is given by [3]

$$T\ln(\tau_{\rm low}) = \Gamma(H,J) = 8J\ell_c - 2|H|(\ell_c^2 - \ell_c + 1)\,, \tag{1}$$

where $\ell_c=\lceil 2J/|H|\rceil$ and $\lceil x\rceil$ denotes the smallest integer not less than x. This result is restricted to $2J/|H|$ not an integer and to $|H|<4J$. This formula shows that the nucleating droplet is a single overturned spin for $2J<|H|<4J$ and an L-shape formed by three overturned spins for $1J<|H|<2J$.

2. The Algorithms

Although in principle traditional Monte Carlo methods could be used to perform the simulation, in practice the computer time required would be prohibitive. Consequently, the MCAMC algorithm is used [7]. Although this algorithm has the same underlying physical dynamic as given above, it implements this dynamic in a fashion that can make the algorithm many orders of magnitude faster than the traditional algorithm. Only a brief outline of the MCAMC algorithm will be given here, see Ref. 7 for details.

The central part of the MCAMC algorithm uses the methodology of Absorbing Markov Chains (AMC). The second calculation method used here also uses AMC. An AMC with q absorbing states and s transient states is given by [8]

$$\mathbf{M} = \begin{pmatrix} \mathbf{I}_{q\times q} & \mathbf{0}_{q\times s} \\ \mathbf{R}_{s\times q} & \mathbf{T}_{s\times s} \end{pmatrix} \tag{2}$$

where the subscripts indicate the submatrix size. The matrix $\mathbf{I}$ is the identity matrix, $\mathbf{0}$ is the zero matrix, $\mathbf{R}$ is the recurrent submatrix with elements equal to the transition probabilities from the transient to the absorbing states, and $\mathbf{T}$ is the transient submatrix with elements equal to the single-time step transition probabilities between transient states. Starting from an initial vector $\vec{v}_{\rm I}^{\rm T}$ that represents the state at $t=0$, the probability of still being in one of the s transient states after m time steps is given by $\vec{v}_{\rm I}^{\rm T}\mathbf{T}^m\vec{e}$, where $\vec{e}$ is the vector with all elements equal to unity. Note that m is an integer, and one time step corresponds to one application of the matrix $\mathbf{M}$. Using a random number r which is uniformly distributed between zero and one, one can obtain a time m for exiting from the s transient states by the solution of the equation

$$\vec{v}_{\rm I}^{\rm T}\mathbf{T}^m\vec{e} < r \le \vec{v}_{\rm I}^{\rm T}\mathbf{T}^{m-1}\vec{e}. \tag{3}$$

Note that m is independent of which of the q absorbing states the system actually does exit to. The elements of the vector of unnormalized probabilities

$$\vec{v}_{\mathrm{I}}^{T}\mathbf{T}_{s\times s}^{m-1}\mathbf{R}_{s\times q} \tag{4}$$

gives the probabilities of exiting to each of the q states. Another random number $\tilde{r}$ can be used to pick which state the system ends up in. Eqs. (3) and (4) are the only equations necessary to utilize AMC in a Monte Carlo simulation. It has been shown [9] that if s=1 and under certain approximations, the MCAMC procedure corresponds to the traditional n-fold way simulation method [10].

If the matrices $\mathbf{T}$ and $\mathbf{R}$ are small enough, the lifetime calculations can be performed exactly. Then it is useful to introduce a matrix called the fundamental matrix, $\mathbf{N}=(\mathbf{I}-\mathbf{T})^{-1}$. The lifetime is then given by $\tau=\vec{v}_{\mathrm{I}}^{T}\mathbf{N}\vec{e}$ [8].

3. Low-Temperature Metastability in the Ising Model

With periodic boundary conditions, the isotropic square-lattice Ising model has n=10 classes of spins. They are ordered as 5 (10) minus the number of nn +1 spins when the spin is +1 (−1) [9,10]. We start with all spins in class 1, and the flip probabilities p_{flip} are given as above by the Metropolis dynamic. For the MCAMC calculations, the final absorbing state is at magnetization equal to zero. For the AMC calculations when the absorbing states are placed at all possible configurations of 3 overturned spins the calculation involves a 6×6 matrix and can be done exactly [11].

The lifetime for $\ell_c=\lceil 2J/|H|\rceil=2$ can be calculated approximately in the low-temperature limit by inverting a 4×4 matrix that has as the transient states only the configurations with one overturned spin, the original configuration, the configuration with two nn overturned spins, and the configuration with a nucleating droplet for ℓ_c=2. The critical droplet in this case is composed to three overturned spins, with the shape that of an L-shape formed by removing one of the squares from a 2×2 square. This gives the lifetime as

$$\tau_2 \approx \frac{1}{p_1 p_2^2}\left[\frac{1}{8}+\frac{1}{4x}\right] \tag{5}$$

where the p_i are the flip probabilities for spins in class i, and x is equal to L^2 times the probability of absorption from a state with the critical droplet to a state with four overturned spins. For Metropolis dynamics x=1, and one has $\tau_2 \approx \frac{3}{8p_1p_2^2}$. Similarly, for ℓ_c=1 one has at low temperatures that $\tau_1\approx\frac{5}{4p_1}$. Consequently, at low temperature the measured lifetime has the form [11]

$$\tau = A\mathrm{e}^{\Gamma/T}\ , \tag{6}$$

where A is a prefactor term independent of the temperature. Specifically, A=5/4 for ℓ_c=1 and A=3/8 for ℓ_c=2.

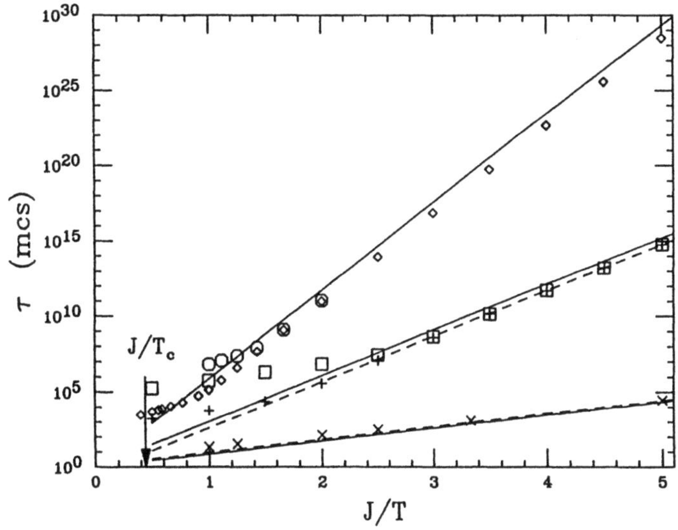

Fig. 1 The average lifetime of the metastable state in units of mcs (Monte Carlo steps) to the state with magnetization zero is shown as a function of the inverse temperature. The simulation points are averages over 1000 escapes for $|H|=3J$ with $L=24$ ($\times$), $|H|=1.5J$ with $L=24$ ($+$) and $L=240$ ($\square$), and $|H|=0.75J$ with $L=24$ ($\diamond$) and $L=240$ ($\circ$). The statistical errors are much smaller than the symbol size. The solid lines are low-temperature predictions without additional prefactors. The dashed lines are low-temperature predictions with a prefactor of 5/4 for $|H|=3J$ and 3/8 for $|H|=1.5J$. The simulation results agree with the low-temperature predictions with an additional prefactor as long as the system is within the single-droplet region for decay, which happens at lower temperatures for larger system sizes.

For more than 3 overturned spins, the AMC must have some absorbing states with a smaller number of overturned spins. In particular, we have performed AMC calculations with 12 transient states, where the largest included cluster has 5 overturned spins. As long as one checks to insure that absorption is extremely improbable in any states with less than 6 overturned spins, the lifetime should be extremely close to the lifetime with a cutoff at 6 overturned spins.

Fig. 1 shows the lifetime as a function of T^{-1} for three values of $|H|$ with three different nucleating droplets. The results are from MCAMC calculations, and seem to not fall on the predicted curve given by Eq. (1). The lifetimes with the predicted prefactors [11] for Eq. (6) are shown as dotted lines in Fig. 1. They agree much better with the data than does Eq. (1).

From Eq. (6) we have

$$\Gamma - T\ln(\tau) = -T\ln(A). \tag{7}$$

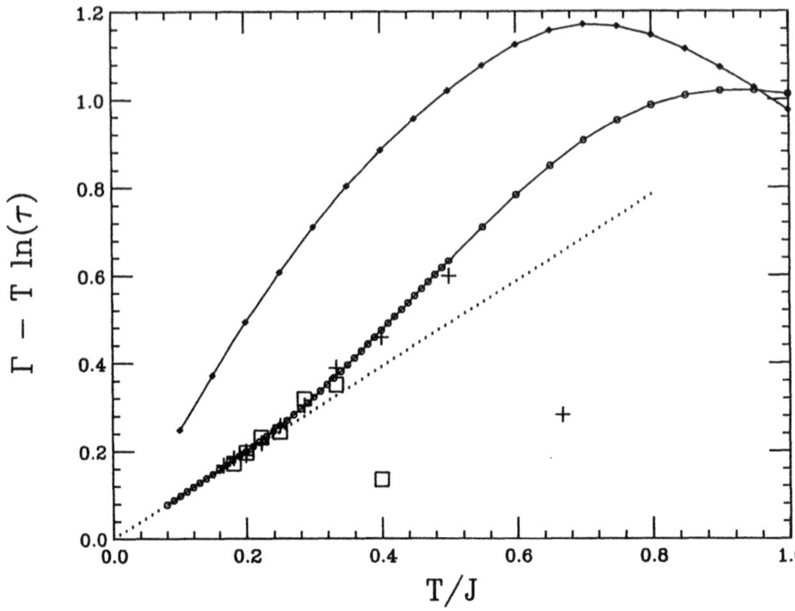

Fig. 2 The difference between the predicted low-temperature result Γ and the temperature times the logarithm of the average lifetime is shown as a function of temperature. The simulation points which are to zero magnetization are for $|H|=1.5J$ and $L=24$ (+) and $L=240$ (□). The statistical errors are much smaller than the symbol size. The point at the highest temperature for both lattice sizes are in the multi-droplet region [1,12,13] and hence the low-temperature predictions are not applicable to them. The absorbing Markov chain results are shown both for absorption to all possible combinations of three overturned spins (◇), and to five overturned spins with additional absorption states for less than five spins (○). The solid lines are guides to the eye. The dashed line has a slope equal to the expected prefactor of 3/8, which is valid for $1J<|H|<2J$. Note that the results all agree with each other at low enough temperature.

A plot of this quantity is shown in Fig. 2 for $|H|=1.50J$. The results from both MCAMC calculations with $L=24$ and $L=240$ agree with the expected results, at least for low enough T once the system is in the single-droplet region and below the dynamic spinodal [12–14]. Also, the AMC calculations that have 12 transient states agree with these results. We see that there is a prefactor A, and the value is in agreement with the predicted value [11] of 3/8.

The AMC results that have all possible combinations of three overturned spins, one configuration of which is the nucleating droplet, does not agree with the results. This is because once the nucleating droplet is formed, at low temperatures there is a probability of 2/3 that the droplet will disappear, whereas this is neglected in this particular AMC calculation.

4. Conclusions

Results for escape from the metastable state for the isotropic square-lattice nn Ising ferromagnet show that for low enough temperature, predictions from low-temperature discrete-droplet theory hold, except there is a temperature independent prefactor which must be included at all non-zero temperatures. This brings into question whether other predicted behavior, such as nucleation through non-equilibrium droplets, will be seen at any finite temperature.

Acknowledgements.
Supported in part by NSF grant #DMR-9520325, and by the Florida State University Supercomputer Computations Research Institute which is partially funded through Contract #DE-FC05-85ER25000 by the U.S. Department of Energy. Also supported by NERSC supercomputer time through the DOE. Useful discussions with P. A. Rikvold are acknowledged.

References

[1] H. L. Richards, S. W. Sides, M. A. Novotny, and P. A. Rikvold, J. Mag. Mag. Mat. **150**, 37 (1995).
[2] H. L. Richards, M. A. Novotny, and P. A. Rikvold, Phys. Rev. B (in press).
[3] E. Jordão Neves and R. H. Schonmann, Commun. Math. Phys. **137**, 209 (1991).
[4] R. Koteckÿ and E. Olivieri, J. Stat. Phys. **70**, 1121 (1993).
[5] R. Koteckÿ and E. Olivieri, J. Stat. Phys. **75**, 409 (1994).
[6] K. Binder, in *Phase Transitions and Critical Phenomena*, eds. C. Domb and M. Green (Academic, London, 1976), Vol. 5B.
[7] M. A. Novotny, Phys. Rev. Lett. **74**, 1 (1995); erratum **75**, 1424 (1995).
[8] M. Iosifescu, *Finite Markov Processes and Their Application*, (John Wiley & Sons, New York, 1980).
[9] M. A. Novotny, Comp. in Phys. **9**, 46 (1995). (There are misprints in Table I of this paper.)
[10] A. B. Bortz, M. H. Kalos, and J. L. Lebowitz, J. Comput. Phys. **17**, 10 (1975).
[11] M. A. Novotny, unpublished.
[12] H. Tomita and S. Miyashita, Phys. Rev. B **46**, 8886 (1992).
[13] P. A. Rikvold, H. Tomita, S. Miyashita, and S. W. Sides, Phys. Rev. E **49**, 5080 (1994).
[14] J. Lee, M. A. Novotny, and P. A. Rikvold, Phys. Rev. E **52**, 356 (1995).

Index of Contributors

Springer Proceedings in Physics

Managing Editor: H. K. V. Lotsch

1 *Fluctuations and Sensitivity in Nonequilibrium Systems*
Editors: W. Horsthemke and D. K. Kondepudi
2 *EXAFS and Near Edge Structure III*
Editors: K. O. Hodgson, B. Hedman, and J. E. Penner-Hahn
3 *Nonlinear Phenomena in Physics*
Editor: F. Claro
4 *Time-Resolved Vibrational Spectroscopy*
Editors: A. Laubereau and M. Stockburger
5 *Physics of Finely Divided Matter*
Editors: N. Boccara and M. Daoud
6 *Aerogels* Editor: J. Fricke
7 *Nonlinear Optics: Materials and Devices*
Editors: C. Flytzanis and J. L. Oudar
8 *Optical Bistability III*
Editors: H. M. Gibbs, P. Mandel, N. Peyghambarian, and S. D. Smith
9 *Ion Formation from Organic Solids (IFOS III)*
Editor: A. Benninghoven
10 *Atomic Transport and Defects in Metals by Neutron Scattering*
Editors: C. Janot, W. Petry, D. Richter, and T. Springer
11 *Biophysical Effects of Steady Magnetic Fields*
Editors: G. Maret, I. Kiepenheuer, and N. Boccara
12 *Quantum Optics IV*
Editors: J. D. Harvey and D. F. Walls
13 *The Physics and Fabrication of Microstructures and Microdevices*
Editors: M. J. Kelly and C. Weisbuch
14 *Magnetic Properties of Low-Dimensional Systems*
Editors: L. M. Falicov and J. L. Morán-López
15 *Gas Flow and Chemical Lasers*
Editor: S. Rosenwaks
16 *Photons and Continuum States of Atoms and Molecules*
Editors: N. K. Rahman, C. Guidotti, and M. Allegrini
17 *Quantum Aspects of Molecular Motions in Solids*
Editors: A. Heidemann, A. Magerl, M. Prager, D. Richter, and T. Springer
18 *Electro-optic and Photorefractive Materials*
Editor: P. Günter
19 *Lasers and Synergetics*
Editors: R. Graham and A. Wunderlin
20 *Primary Processes in Photobiology*
Editor: T. Kobayashi
21 *Physics of Amphiphilic Layers*
Editors: J. Meunier, D. Langevin, and N. Boccara
22 *Semiconductor Interfaces: Formation and Properties*
Editors: G. Le Lay, J. Derrien, and N. Boccara
23 *Magnetic Excitations and Fluctuations II*
Editors: U. Balucani, S. W. Lovesey, M. G. Rasetti, and V. Tognetti
24 *Recent Topics in Theoretical Physics*
Editor: H. Takayama
25 *Excitons in Confined Systems*
Editors: R. Del Sole, A. D'Andrea, and A. Lapiccirella
26 *The Elementary Structure of Matter*
Editors: J.-M. Richard, E. Aslanides, and N. Boccara
27 *Competing Interactions and Microstructures: Statics and Dynamics*
Editors: R. LeSar, A. Bishop, and R. Heffner
28 *Anderson Localization*
Editors: T. Ando and H. Fukuyama
29 *Polymer Motion in Dense Systems*
Editors: D. Richter and T. Springer
30 *Short-Wavelength Lasers and Their Applications*
Editor: C. Yamanaka
31 *Quantum String Theory*
Editors: N. Kawamoto and T. Kugo
32 *Universalities in Condensed-Matter*
Editors: R. Jullien, L. Peliti, R. Rammal, and N. Boccara
33 *Computer Simulation Studies in Condensed-Matter Physics: Recent Developments*
Editors: D. P. Landau, K. K. Mon, and H.-B. Schüttler
34 *Amorphous and Crystalline Silicon Carbide and Related Materials*
Editors: G. L. Harris and C. Y.-W. Yang
35 *Polycrystalline Semiconductors: Grain Boundaries and Interfaces*
Editors: H. J. Möller, H. P. Strunk, and J. H. Werner
36 *Nonlinear Optics of Organics and Semiconductors*
Editor: T. Kobayashi
37 *Dynamics of Disordered Materials*
Editors: D. Richter, A. J. Dianoux, W. Petry, and J. Teixeira
38 *Electroluminescence*
Editors: S. Shionoya and H. Kobayashi
39 *Disorder and Nonlinearity*
Editors: A. R. Bishop, D. K. Campbell, and S. Pnevmatikos
40 *Static and Dynamic Properties of Liquids*
Editors: M. Davidović and A. K. Soper
41 *Quantum Optics V*
Editors: J. D. Harvey and D. F. Walls
42 *Molecular Basis of Polymer Networks*
Editors: A. Baumgärtner and C. E. Picot
43 *Amorphous and Crystalline Silicon Carbide II: Recent Developments*
Editors: M. M. Rahman, C. Y.-W. Yang, and G. L. Harris
44 *Optical Fiber Sensors*
Editors: H. J. Arditty, J. P. Dakin, and R. Th. Kersten
45 *Computer Simulation Studies in Condensed-Matter Physics II: New Directions*
Editors: D. P. Landau, K. K. Mon, and H.-B. Schüttler

MIX
Papier aus verantwortungsvollen Quellen
Paper from responsible sources
FSC® C105338

If you have any concerns about our products,
you can contact us on
ProductSafety@springernature.com

In case Publisher is established outside the EU,
the EU authorized representative is:
Springer Nature Customer Service Center GmbH
Europaplatz 3, 69115 Heidelberg, Germany

Printed by Libri Plureos GmbH
in Hamburg, Germany